T5-AGX-370

GEOLOGIA

This is a volume in the Arno Press collection

History of Geology

See last pages of this volume
for a complete list of titles

GEOLOGIA

Or

A Discourse Concerning the Earth Before the Deluge

Erasmus Warren

ARNO PRESS

A New York Times Company

New York / 1978

Editorial Supervision: ANDREA HICKS

Reprint Edition 1978 by Arno Press Inc.

HISTORY OF GEOLOGY
ISBN for complete set: 0-405-10429-4
See last pages of this volume for titles.

Manufactured in the United States of America

Publisher's Note: This book has been reproduced from the best available copy.

Library of Congress Cataloging in Publication Data

Warren, Erasmus.
Geologia : or, A discourse concerning the earth before the deluge.

(History of geology)
Reprint of the 1690 ed. printed for R. Chiswell, London.
1. Geology--Early works to 1800. 2. Deluge.
3. Burnet, Thomas, 1635?-1715. Telluris theoria sacra.
I. Title. II. Series.
QE25.W37 1977 550 77-6546
ISBN 0-405-10470-7

GEOLOGIA:

OR, A

DISCOURSE

Concerning the

EARTH before the DELUGE.

WHEREIN

The FORM and PROPERTIES aſcribed to it,

In a Book intituled

The Theory of the Earth,

Are Excepted againſt:

And it is made appear,

That the DISSOLUTION of that Earth was not the Cauſe of the Univerſal Flood.

ALSO

A New Explication of that Flood is attempted.

By *ERASMUS WARREN*, Rector of *Worlington*, in *Suffolk*.

גם את־העולם נתן בלבם;

ECCLESIAST. iii. 11.

Et Mundum tradidit Diſputationi eorum.

LONDON,

Printed for R. Chiſwell, at the *Roſe* and *Crown* in St. *Paul's* Church-Yard. MDCXC.

TO THE
READER.

HAving peruſed the Book called *The Theory of the Earth*: conſidering it ſimply and abſtractedly in it ſelf, as a Philoſophic *Scheme* or repreſentation of things; I found it a Treatiſe, not unworthy of the ingenious Author of it. Though ſo it was not without its σφάλματα, or Defects. But then taking it as it relates to the Doctrine of the *Bible*, and ſo bears the Title, * *Sacred*; I thought it liable to ſeveral *Exceptions*.

* *Hanc Theoriam Sacram appello, cum Telluris Phyſiologiam communem non reſpiciat; ſed majores mundi noſtri viciſſitudines, quarum meminit Sacra Scriptura.* Præfat. ad Lectorem.

Some of theſe I determin'd to ſet down forthwith, and in a Letter tranſmit them to the learned *Theoriſt*. But my Pen growing warm, quite out-run the bounds of my firſt Intentions; and forcing me to alter the Method I had begun, carry'd things on to this length, and drew them up in this Form.

My Design is only an humble Proposal of some few *Exceptions* against the *Essentials* of the *Theory*. And I as humbly beg, that they may not be mistaken, as to their *Rise*; nor misconstrued, as to their *End*. They flow but from *Affection* to *Truth*; and are directed to her *Vindication*. Let none therefore think them off-springs of a narrow mind, or issues of a captious disputatious Spirit; much less of a stingy Picque against Philosophy, to which, as I owe all becoming veneration, so I shall ever duly pay it.

Nor have I ingaged with the Theory at all because it is *New*; but because it is *False*. For all that is *true*, must needs be *ancient*: only the *Discovery* of some truth may be *New*. But then every such discovery of important truth, is highly to be valued and applauded: To be welcom'd into the World with thanks and joy; and entertain'd with reverence and a sweet reception. Yea as every such Discovery of weighty Truth, ought to be receiv'd with great kindness and respect; so that happy Person, who makes the Discovery, ought to be exceedingly honoured too, and lookt upon as deserving a Coronet and a Monument. And for

for my own part, I had much rather assist with my hands in fitting up both; than write one word or syllable with my Pen, to hinder him of either.

Again therefore I heartily profess my Scope, to be nothing else but a *Vindication* of *Truth*; unless I shall add, and of *Religion* also. For though I am far from that temper, of being * *alarm'd at the proposal of every new Theory, as if all Religion were falling about our ears*: yet I am sensible the *Theorist* has assaulted *Religion*, and that in the very *foundation* of it. And therefore he must not blame me, if I have taken the *alarm* (to use his own word) when he was pleased in such a manner to give it. And truly should not some of us have been awakened by it, considering how loud he rings it in our ears; he might well have concluded, we were too fast asleep.

* In the Preface to the Eng. Theory.

When the fourteen † Books of *Numa Pompilius*, that ancient and famous King of *Rome*, were found in the Earth in a Chest of Stone; and being taken out were perused by several: at last, upon the Prætor *Petilius*'s report, that they contained *pleraque dissolvendarum Religionum*, *many things* tending *to the undoing of Religion* (at least in some Ceremonies or appendages of it) they were condemned by the

† Liv. Hist. Dec. 4. lib. 10.

Senate

Senate——An Argument of their tenderneſs and due concern for the Religion they had, though but a mean one. Now allowing our regard for Religion, to exceed that of the *Romans* (as in reaſon it ſhould) but as much as *our* Religion exceedeth *theirs*: and how deeply muſt we reſent (eſpecially thoſe of us in Holy Orders) even the ſmalleſt injuries done unto it? But then when Books come forth too like to *Numa*'s, the Contents whereof ſtrike at *Religion*; the leaſt we can do, is to complain of the abuſe, and endeavour meekly to confute them.

And that *the Theory of the Earth* does ſtrike at Religion, and *aſſault* it (as I ſaid) in the very *Foundation* of it; is but too evident. For in ſeveral things (as will appear in our Diſcourſe) it contradicts *Scripture*; and by * *too poſitive* aſſerting the truth of its own Theorems, makes *that* to be falſe; upon which our Religion is founded. And to aſſert ſuch things poſitively, as imply Scripture to be falſe in any Periods of it; muſt be of very pernicious conſequence. For if it fails in *ſome* inſtances, it may do ſo in *many*: and that which renders it ſuſpected in *part*, will impeach the credit of the *whole*.

* Read the 17th Chapter of this Diſcourſe.

Let

Let it be noted therefore that the Diſpute here, is not meerly whether the World we live in, be the ſame *now*, as it was of *old* before the Flood: or whether there be not as much difference betwixt its *primæval*, and its *preſent* State; as betwixt a goodly Structure, when ſtanding in its glory, and groveling in its ruines: but (which is far more material) whether ſome ſacred and *revealed Truths*; or gay, but groundleſs *Philoſophic Phancies*; ſhall be preferred.

The Book has lain by in Manuſcript a great while. Why it did ſo, is well known to *ſome* good Men; and I need not trouble *all* with the Reaſons of it. But when none (as I could learn) were diſpos'd to write better; I let it come abroad.

In it, I have not (to ſpeak in the *Theoriſt's* language) * *made Judgment or Cenſure of* his Hypotheſis, *upon general preſumptions and prejudices, nor according to the temper and model of my own ſpirit; but* (I think) *according to reaſon.* And that I might not *impoſe upon my ſelf or others, have laid aſide that lazy and fallacious method of cenſuring by the lump,* and endeavoured *to bring things cloſe to the Teſt of* true *and* falſe, *to explicit proof and evidence. And whoſoever,* ſays he, *makes ſuch Objections*

* In the Preface to the Eng. Theory.

againſt

against an Hypothesis, *hath a* **Right** *to be heard.* This *Right* therefore, so far as it is mine, and I may lawfully do it; I now challenge.

To conclude. Whereas I have endeavoured to explain the Universal Deluge, in a new and unusual way; I would by no means be thought to ground upon it, as certainly true: but only to show that another way of opening and unfolding that intricate *Phænomenon*, may be found out, as *plausible* or *approvable* as that which the *Theory* goes in. And truly for my own part, I am much of the Opinion of a very learned Friend of mine (a great ornament both to the University and the Faculty he is of) who upon perusal of this Book in Manuscript, wrote this to me among other things: *Though we have* Moses, *yet I believe we must stay for* Elias, *to make out to us, the true Philosophical* modus *of the Creation, and Deluge.*

THE

THE CONTENTS.

CHAP. I.

CHAP. II.

CHAP. III.

CHAP. IV.

CHAP.

The CONTENTS.

CHAP. V.

CHAP. VI.

CHAP. VII.

CHAP. VIII.

 time.

CHAP. IX.

CHAP. X.

CHAP. XI.

CHAP.

The CONTENTS.

CHAP. XII.

CHAP. XIII.

CHAP. XIV.

CHAP.

ERRATA.

PAge 13. Line 10. Read *incorruption*. p. 48. l. 3. after *that*, insert *were in it*. p. 58. l. 3. r. *host of*. p. 59. l. 31. r. המדרש. p. 60. l. 26. r. למסורת. p. 75. l. 13. r. *and*. p. 95. l. 1. r. *professed*. p. 98. l. 23. r. אינו. p. 109. l. 21. r. *canales*. p. 116. l. 7. r. *Miles*. p. 127 l. 9. r. *Brahe*. p. 129. l. 25. r. *descry*. p. 145. l. 28. r *inartificial*. p. 233. l. 1. r. *grow*. l. 5. blot out *so*. p. 255. l. 1. r. *just*. l. 30. r. *its*. p. 282. l. 14 r. *Crops*. p. 289. l. 21. after *land*, insert, *excepting the Red Sea*. p. 306. in the last line of the Margent, r. והומות. p. 307. l. 17. blot out *a*. p. 321. l. 25. r. *two hundred*. p. 324. l. 17. after *in*, r. *answered*. p. 333. l. 2. r. *about*. p. 345. l. 25. r. *hideous*. The Parentheses, p. 289. l. 15, 16, 17, 18. and p. 290. l. 2, 3. should have been left out. Some *mispointings* also must be noted.

GEOLOGIA:

GEOLOGIA:

OR, A

DISCOURSE

Concerning the

EARTH

Before the Flood.

CHAP. I.

1. *The great* Uſefulneſs *of* Natural *Philoſophy.* 2. *In proving there* is *a God.* 3. *In acquainting us with His* Nature. 4. *In aſſerting a* Providence. 5. *In excluding* Idolatry. 6. *In vindicating the* Goſpel *in* ſeveral *Points.* 7. *As the Immortality of the Soul.* 8. *The Reſurrection of the Body.* 9. *The Conflagration of the World.* 10. *And the endleſs fiery Torments of the Damned.* 11. *Philoſophy is uſeful alſo as to* Divinity. 12. *And like to* flouriſh. 13. *Caution againſt* abuſing *it.* 14. *Which is done, either by* ſpeaking, *or* thinking ſlightly *of it.* 15. *Or by Setting*

ting it too low *in its Operations.* 16. *Or else by Raising it too* high. 17. *Which is the fault of* The Theory of the Earth. 18. *A* Character *of it.* 19. *The* Occasion *of this Discourse against it.* 20. *Together with its* Method. 21. *This Chapter, an* Introduction *to the Discourse.*

1. IT is a memorable and worthy Saying (for a Heathen) of † *Simplicius*; *Philosophy is the greatest Gift that ever GOD bestowed upon Men.* And were it restrained to *Natural Philosophy* alone, there would be much truth in that Assertion of his. For it serves our interests with a mighty efficacy, and is highly conducive to our benefit; not only many, but innumerable ways. Thus it exalts our Minds, and inlarges our Understandings, and fills them with rich and invaluable Notions. It elevates our flat and groveling Souls, and makes them at once to look up, and look high. It disinthralls our Judgments, inslav'd to Sense, and weak Speculations; and swells our shrivel'd narrow Thoughts, into wide, and generous, and comprehensive Theories. It wipes the dust of Ignorance, and dimness of Prejudice out of our Eyes; and inables us not only to see Nature's Beauty, but duely to admire it. Yea, in a short time, it turns our Admiration into studious Industry; and of passionate Lovers of Nature's Perfections, makes us curious and painful Searchers into her Mysteries. And here new Discoveries bring fresh * Delights; and our intellectual Satisfactions, do more than compensate our most tiresom Disquisitions. For the Mind being weighed down with

† Τὸ γὰρ μέγιστον τῶν ἐκ Θεοῦ δωρηθέντων, φιλοσοφία ἐστί. *In Epict. Ench. cap.* 29.

* *Voluptas quam percipimus ex intuitu rerum quas oculi cernunt, minime æquiparanda est cum illa quam adfert notitia illarum quas philosophando invenimus.* Des Cartes in Præfat. ad Princip.

the

the luggage of the Body, and bound faſt, as with Chains, in the ſtraitneſſes of it; Philoſophy relieves it (ſays a great (*a*) Man) by giving it a fair Proſpect of the things of Nature, and lifting it up from Earthly, to Divine Concerns. To take cognoſcence of which, while it ſallies out, it recovers a kind of liberty; and breaking looſe, in ſome ſenſe, from the uneaſie preſſure and confinement it ſuffers, is refreſhed with the ſurvey and ſtudy of the Heavens.

(*a*) *Corpus hoc, animi pondus ac pœna eſt: premente illo urgetur: in vinculis eſt, niſi acceſſit philoſophia: & illum reſpirare rerum naturæ ſpectaculo juſſit, & à terrenis dimiſit ad divina. Hæc libertas ejus eſt, hæc evagatio; ſubducit interim ſe cuſtodiæ qua tenetur, & cœlo reficitur.* Sen. Ep. 65.

The learned (*b*) Father flies higher ſtill, though not in the leaſt above the Mark. For he makes Philoſophy *profitable for Godlineſs, to ſuch as fetch Faith from Demonſtration.* And ſays, *That if it does not comprehend the vaſtneſs of Truth, nor is able to perform the Commandments of the* LORD; *yet it makes way for the moſt royal Doctrine.* And therefore he would have *all* (not excuſing very (*c*) *Women*) to mind Philoſophy. And argues, That none who are (*d*) *young* ſhould defer it; and that none who are *old* ſhould be weary of it; *becauſe no Man is too young, nor yet too old, to get a ſound Mind.* And then adds, *He that ſays 'tis too ſoon, or too late, to ſtudy Philoſophy; is juſt like him who ſays, it is too ſoon, or too late, to be bleſſed.* And that Philoſophy ſhould contribute towards Mens Bleſſedneſs, we need not wonder; when (as he ſays in

(*b*) Εἰ μὴν μὴ καταλαμβάνει τὸ μέγεθος τῆς ἀληθείας, ἔτι δὲ ἐξασθενεῖ πράττειν τὰς κυριακὰς ἐντολάς, ἀλλ' οὖν γε προκατασκευάζει τὴν ὁδὸν τῇ βασιλικωτάτῃ διδασκαλίᾳ. *Clem. Alex. Strom. lib.* 1.

(*c*) Φιλοσοφητέον καὶ ταῖς γυναιξίν. *Strom. lib.* 4.

(*d*) Μήτε νέος τις ὤν, μελλέτω φιλοσοφεῖν, μήτε γέρων ὑπάρχων, κοπιάτω φιλοσοφῶν. οὔτε γὰρ ἄωρος οὐδείς ἐστιν, οὔτε πάρωρος πρὸς τὸ κατὰ ψυχὴν ὑγιαίνειν. ὁ δὲ λέγων, μήπω τοῦ φιλοσοφεῖν ὑπάρχειν ὥραν, ἢ παρεληλυθέναι τὴν ὥραν, ὅμοιός ἐστι τῷ λέγοντι, πρὸς εὐδαιμονίαν ἢ μήπω παρεῖναι τὴν ὥραν, ἢ μηκέτ' εἶναι τὴν ὥραν. *ibid.*

in (*a*) another place) *it does before-hand purge and prepare the Soul to receive the Faith, upon which the truth builds Knowledge.* And albeit in theſe Expreſſions, he might not mean *Natural* Philoſophy only; yet ſpeaking all along of the *Greek* Philoſophy in general, he cannot be ſuppoſed to exclude that neither. Which indeed does very much qualifie and diſpoſe us for true Religion; and is rarely inſtrumental to improve and advance it.

(*a*) Προκαθαίρει καὶ προεθίζει τὴν ψυχὴν εἰς παραδοχὴν πίστεως, ἐφ' ᾗ τὴν γνῶσιν ἐποικοδομεῖ ἡ ἀλήθεια. *Strom. lib.7.*

To make out this fully, how uſeful and ſerviceable Philoſophy is, to promote Religion; would require a whole Volume. Let me only touch upon a few Particulars.

2. Firſt, it is uſeful, *To prove, There is a GOD.* Of all the Fundamentals of Religion, this is the chief. Yet if Philoſophy did not lend us ſome Topicks, from whence we might fetch Arguments, to evince and confirm the Exiſtence of a DEITY; all that we could ſay, would be too cold and languid to confute the Atheiſt. But when that diſcovers an abſolute neceſſity of a Firſt Cauſe, and of a Firſt Mover; and of an infinitely Wiſe and Powerful Creator of the Univerſe, and of as infinitely Wiſe and Powerful a Governour of the ſame: Or elſe on the other hand, ſhews a neceſſity of Deifying the World itſelf, by beſtowing Godlike Attributes upon it; and of granting *Self-movency*, *Life* and *Underſtanding* to Matter; with other moſt notorious and numberleſs Abſurdities: Then he muſt either openly confeſs there is a GOD, or with ſilence ſubmit to a Belief of his Being.

And what a conſiderable ſtroke Philoſophy has in proving there is a GOD, *Plutarch* fairly gives us to underſtand; where he declares, (*b*) *That the firſt*

(*b*) Ἔσχον δὲ ἔννοιαν τούτου πρῶτον μὲν ἀπὸ τοῦ κάλλους τῶν ἐμφαινομένων προσλαμβάνοντες. *De Placit. Philoſ. lib. 1. cap. 6.*

first Notion Men had of him, they took from the beauty of the aspectable things. And a little after, (a) *They had the knowledge of a* G O D, *from the Stars which appeared, while they beheld the great harmony they caused, and how orderly they made Day and Night, and Winter and Summer.* To which agrees what we read, *Wisd.* 13.3,4.

(a) Θεοῦ ἔννοιαν ἔχειν ἀπὸ τῶν φαινομένων ἀστέρων, ὁρῶντες τούτους μεγάλης συμφωνίας ὄντας αἰτίους, καὶ τεταγμένας ἡμέραν τε καὶ νύκτα, χειμῶνά τε καὶ θέρος. *ibid.*

3. Secondly, *To acquaint us with the Nature of the* DEITY. For what to make of his Immaterial or Spiritual Essence; of his Necessary and Self-existence; of his Ubiquity or Omnipresence; of a TRINITY in Unity; or Three distinct Persons, in one and the same undivided Nature, and common essential Substance, *&c.* we should be utterly at a loss, were it not for Philosophy. Not that Philosophy can enable us to look to the Centre of God's Perfections neither, and throughly to understand him; that's impossible. For he is nothing else but GLORY and GREATNESS. And such is the Brightness of the one, and the Immenseness of the other, yea, the Infiniteness of both, that no created Capacities, with all the helps they can possibly get, shall attain to a clear and full knowledge of him. Yea, much of the felicity of the Eternal State, seems to lye in this; That as we shall always see more and more of GOD, so we shall never be able to discern *all*: But shall incessantly be entertained with fresh Perceptions of new Delights, arising from fresh Apprehensions, and new Discoveries of the incomprehensible Goodness and Beauties of the Divinity. Which Apprehensions will be so very clear, and the Oblectations issuing from them so high and strong, in reiterated rapturous Vibrations in the Soul, that we shall be strangely overflowed, and

and as it were quite swallowed up of endless and most beatifying Satisfactions. And O amazing Bliss and Happiness indeed, where we shall ever be sinking deeper and deeper still in an abyss of intellectual Joy and Sweetness! This will make our condition a boundless Ocean of transporting Pleasures; as GOD's Nature is a like Ocean of Divinest Excellencies. But then if GOD be too Glorious and Great to be perfectly understood by us in the Mansions above (where the dormant Powers and Faculties of our Souls (which perhaps are many) shall be all awakened into lively Actings) how much less can Philosophy help us to understand him compleatly, here in these lower Regions? Yet as Men may see more with a good Perspective, than they can do without it; so we may better acquaint our selves with the Nature of GOD, being assisted by Philosophy, than if we had it not. And the truth is, even the very *Word* and *Works* of GOD, the two most informing things we have, the most apt and able to lead us into competent knowledge of His MAJESTY, without Philosophy, are neither of them to be rightly or tolerably understood in innumerable Instances.

4. Thirdly, *To assert a Providence, and the free use of our Faculties. Astra regunt homines, Stars govern Men,* has taken great place in the World: Insomuch that a Mans *Stars*, and his *Destiny*, have been Terms equivalent. As free Agents as we are, they have been thought to controll us; to be Disposers of our Lot, and Dispensers of our Portion, and the Masters of our Fortune. To incline our Minds, and to sway our Wills; to determin our Motions, and over-rule our Actions; staking us down as it were in all our Proceedings by irresistible force, or tying us up to fatal necessity.

necessity. A Persuasion most strange, and also as false. It might come from hence. The Stars were observed to be very numerous, and for their largeness, most considerable Bodies. And therefore for them only to shine in the Night-time, and that with so pitiful a light, as every little Fog or Cloud can obscure, might well be thought too mean a work for them, at least for their whole Employment. Whereupon to intitle them to another Task, more noble and suitable, they were fancied to have Mankind committed to their charge, to be ruled by them with an absolute Regency. But whence sprang this Mistake, of the Stars Superintendency and Sovereign Dominion, save merely from the want of sound Philosophy? For as many as are tolerably vers'd in that, know they can have no such Influence or Empire; but are wholly incapable of exercising a Regiment of that Nature over us. And do know as well likewise to what other great Ends or Uses they may serve, most worthy of themselves; much better than to such an impossible Jurisdiction, as has been wildly and unreasonably attributed to them.

5. Fourthly, *To exclude Idolatry.* The most general and constant Idol among Men, has been the Sun. And how came he to be so? Why, from his Motion, he might probably be reputed an Animal: And from his regular Motion, he might as well be thought to have Understanding. And then from these Notions or Persuasions concerning him, Men might easily ascend to an higher yet, and fancy him a Being of Divine Perfections; as *Cardan* and *Vaninus* did. And so it would be natural to honour him as a GOD, by paying Divine Worship to him. Yea, a long and prevailing Doctrine it has been, That the Sun (though he

he was not a rational and underſtanding Creature himſelf) was actuated and directed by a Spirit or *Intelligence*. And this ſeems an Opinion, but one remove from the other; and might give too juſt occaſion and encouragement to ſuperſtitious Adoration of the Sun; if not open a wide door whereat it might enter. But then true Philoſophy interpoſing here, makes ſo full a diſcovery of his Nature, as may throughly aboliſh and for ever deſtroy all grounds and reaſons of Idolatrous Practices in reference to him. Though through want of ſuch Philoſophy at *Athens*, *Anaxagoras* fell under a double misfortune; being at once both fin'd and baniſh'd, for calling the Sun a meer Globe of Fire.

The like alſo is applicable to the Stars, the Moon, and the reſt of the Planets; which too often, and with too many, have been reputed Deities, and treated accordingly. And therefore *Philo* ſays, they were (a) *of old*, αἰσθητῶν Θεῶν εὐδαίμων νομισθεὶς στρατὸς, *the reputed bleſſed Hoſt of ſenſible* GODS.

(a) *Lib. de mund. incorrupt.*

6. Fifthly, *To vindicate the Goſpel.* In illuſtrating, that is, and clearing it up, in ſome of its moſt conſiderable Points or Articles. In it are delivered the moſt high and important things that can be; and they ſeem to be as difficult in their Proof, as they are lofty and momentous in their Nature and Accompliſhment. But from Philoſophy they receive good Light and Confirmation.

7. Of this kind, in the firſt place, is, *The Immortality of the Soul.* The certain truth or reality of which, is abundantly evidenc'd by plenty of Arguments from the ſeveral *Places* of Philoſophy. Thus, it teaches (for Inſtance) that the Soul is a ſpiritual or immaterial

terial ſubſtance: and ſo, diſtinct from the Body, and independent upon it. Which Propoſition alone, throughly confirmed, is ſufficient to eſtabliſh her Immortality; and that ſtrongly and impregnably, againſt all the Cavils and perverſe Objections, that wanton Wits of captious *Somatiſts* can raiſe in contempt, or (as they imagine) in confutation of it. And that the Soul is a purely ſpiritual Being, and ſo, quite different from, and exalted above the Nature of Body; even then when ſhe animates it, and lives in cloſeſt conjunction with it: Philoſophy gives us clearly to underſtand, from her POWERS or FACULTIES, and from her OPERATIONS. I mean while it inables us to apprehend and judge, that they are too lofty and excellent, too regular and active (as being vital and intellectual) to be in any meaſure compatible to meer corporeity; though of never ſo ſublimate and refin'd a conſiſtency.

Thus, we know, the Soul can move and perceive, and underſtand and conſider, and reaſon and conclude, and deliberate and determine, and chooſe and refuſe. That ſhe is capable of framing univerſal Propoſitions, and of apprehending ſpiritual and abſtracted Eſſences, and the Numbers, and Notions, and Ideaes of things. That ſhe is able to correct the ſeeming repreſentations of Senſe and Fantaſie; and when Objects appear in ſuch Diſtance, or Figure, or Colour, or Magnitude: to think and infer that they muſt be otherwiſe. That ſhe has power to ſtrive with the Bodily Appetites; and after many, and long, and vehement Colluctations, to maſter and ſubdue them. She going one way, by vertue of that free and innate Rule, which is founded and deeply radicated in her ſelf: while ſly Propenſities and Inclinations flowing from inferiour ignoble Paſſions;

C do

do dispose her *to*, and would impetuously hurry her on *in* another. Where though compliance with the motions she feels, would be far more pleasing to her at present, than stout resistance; yet she bravely stands out, because that in her prospect of future consequents, she is sensible, yielding would turn to her prejudice. Yea, she cannot only encounter and fight against the Body, and maintain both stiff and lasting conflicts with its brutish affections; but moreover delight in the Combates while they hold, and when they are past, rejoice in the Victory.

Now all these Powers and Workings of the Soul, in the Accounts of Philosophy, are Arguments and Indications of her Incorporealness, or distinctness from the Body. And if we look upon them, as supervening *to*, or falling in *with* the Evangelical Doctrine of her Immortality; the Article will thereby be eternally settled, never to be shaken, much less to be ruined, by the most powerful Assaults and furious Batteries, that Atheistical Sophistry can make against it.

We are all mortal, and he that lives longest, must die at last. But when he does so, some are apt to surmise, *quòd totus moritur*, that he dies wholly and finally, without or beyond all hope and remedy. And truly were the Soul no more than what many would make her; namely, *Crasis, Harmonia, vel Modus corporis*; *a Temperament, Harmony, or Modification of the Body*; his fate might well be supposed such, because indeed it could be no other. But sound Philosophy, I say, makes it very evident, that the Soul is a substance distinct from the Body. That there is as real, and as great a difference between them, as there is between the Prisoner and his Chains; or the House and its Inhabitant. And therefore several Philosophers

phers (and *Socrates* for one) have taught us to call this Body, τὸ ψυχῆς οἰκητήριον, *the habitation of the Soul*; yea, φυλακὴ, καὶ τάφος, her moveable *Prison*, and her living *Sepulchre*. And the *Essenes* (followers of *Pythagoras*) reckoned, as *Josephus* informs us, ὥσπερ ἐκ τῶν δεσμῶν, &c. that to go out of the Body, was *like an inlargement from Fetters*. And this distinctness of the Soul from the Body, does aloud proclaim her independence upon it; and shows her to have a vital Power in her self, inabling her to subsist when separated from it.

Yet (to shut up the Point) none need marvel that the excellent Philosopher now mentioned, I mean *Socrates*, should leave the World with a seeming mistrust of this great Truth, the Immortality of the Soul, (with an, εἴπερ γε τὰ λεγόμενα ἀληθῆ ἐστι, *IF SO BE the things spoken be true.*) For the wise Man's condition, at the present, was exceeding black and dark; and the gloomy lowring circumstances he was in, might well cloud his mind, and dull his thoughts, and so depress and sink his Spirit; as to hinder his quick apprehension of things. And though he was a very considerable Philosopher in his time, yet the Gospel which *brought life and * immortality to light*, * 2 Tim. 1. 10.
had not then visited the World: nor did Philosophy it self, then shine so bright by much, as now it does. All which rightly weighed, instead of wondring at his diffidence, we may rather be amazed at his Courage and Gallantry. For they who shall read, what *Plato* relates (in his *Apology*) concerning him, may well be surprized, not to say, astonished at his exemplary fortitude; and the generous and incomparable greatness of his Spirit, at his dying hour.

8. A second grand Doctrine that Philosophy vindicates, by contributing towards the elucidation of it, is, *The Resurrection of the Body*, which was never explicitly recommended to the World, till the Gospel was promulgated. And therefore is fitly called by *Tertullian*, *fiducia Christianorum*, the proper or peculiar *confidence of Christians*.

This the Atheist can seldom either read, or hear of, but he is ready to smile, if not to insult. He huggs it as a dear and most useful notion; As a notion which he thinks he can form into an Engin, and that of such mighty and irresistible force, as shall be able to batter down the whole Fabrick of the Christian Religion. That Mens Bodies should die, and be buried, and rot in their Graves, and grow up into Grass; and this Grass be eaten by Beasts, and those Beasts be eaten by Men, and those Men be eaten by Canibals, and those Canibals devoured by Worms, and those Worms turned to Dust, and that Dust quite dissipated and lost; and yet those Bodies rise again, hundreds or thousands of Years after: How strange and impossible must this be? How vain and foolish the Religion that teaches it? How fond and silly the People that believe it? But Philosophy helps to make out the mystery.

For that teaches, *That no MATTER does perish*; that none of it is annihilated or utterly lost. Whereupon it follows, That when our Bodies are interred, and reduced to Earth, or else burned, and for the most part evaporate into the Air, or the like: they suffer but a dissolution and dispersion at worst. And therefore they may still be capable of being gathered up in their widest dissipations, and of being made into living Bodies again. Provided there be but (as there is) a Power in the World, that can work them to re-unions, by by consigning and adjusting their numberless Particles, into their respective integral Masses.

Philosophy teaches also, *That all Matter in being was once the* same ; *and that it is so still, only it is diversified into several kinds, by various Modifications.* Whence it follows, That (according to the account we have of the Resurrection) a *Terrestrial* Body, may become a *Celestial* one; and a *Natural* Body, may become a *Spiritual* one; this *Corruptible,* may put on *Incorruptible* ; and this *Mortal,* put on *Immortality* ; and yet be the very *identical* Body that it formerly was ; the same in *Substance,* only different in *Qualities.* As Gold, is the same Gold, when tried and refined ; only more pure than it was before.

Farther, the same Philosophy instructs us, *That the most of that Food, which Men take, does not come to Assimilation, or absolute conversion into the substance of their Bodies* : yea, *that very little of it does so.* Whence it follows again, That the Body of a Man by being eaten of Cannibals (though it should be eaten over and over) needs not for *that,* be hindred from rising again. For, excepting some slender portions of it, it slips through the eaters ; and however by Concoction it be turned into Chyle, yet it is not carried on to a full transmutation, and converted into Flesh. So that the greatest part of the devoured Carkass, is only altered in it self, by passing through the Vessels of them that eat it ; but no way appropriate to another individual. And the remains of this macerated changed Carcass (that never ran into the Composition of others) after all its mutations, being turned at last (to use *Tertullian's* word) into *the Substance of Eternity* ; shall thereby be fitted to re-imbody that Soul, which formerly wore

wore it. Yet ſtill it will be the *ſame* Body (in a very good (let me ſay) in a *Goſpel* ſenſe) that it was when it died; foraſmuch as it muſt conſiſt of the ſame matter which did then conſtitute it. I, and we may ſtill call it *Fleſh* too, becauſe it really is ſo. Only inſtead of *Corruptible* Fleſh (which
*1 Cor. 15. 50. not only *ſhall not*, but ὐ δύναται, * cannot *inherit the kingdom of* GOD; we muſt allow it to be *Glorious* (let it not ſeem abſurd, if I add) *Spiritual* Fleſh. In a word, it ſhall be ſuch Fleſh, as our BLESSED REDEEMER's is now in Heaven. And therefore we may remember, he has promiſed, by a famous Apoſtle; when he ſhall appear again,
* Phil. 3. 21. * μετασχηματίζειν σῶμα τ̃ ταπεινώσεως ἡμῶν, *to transform the Body of our humiliation*; and to make it σύμμορφον τῷ σώματι τῆς δόξης αὐτῦ, *conform to the Body of his Glory.* Where, μετασχηματίζειν, does plainly ſignifie, That the Bodies of Holy Men, at the Reſurrection, are not to be changed ἐν ὐσίᾳ, *in ſubſtance*; but ἐν σχήματι, in *Form* only. The *matter* of them is ſtill to be the ſame; only the *Modes* of them are to be altered, and they muſt put on new, and better Accidents. The variation they ſuffer, ſhall be in the *Scheme* or Habit; not in the *Eſſence* of their Bodies.

Two lively *Specimens*, or Pledges of this admirable change, which at laſt ſhall happen to the Bodies of the Righteous, to their great improvement in circumſtantials; are given us in the inſpired writings. Which, if duly conſidered, may ſerve to ſtrengthen our Faith, and make us firm and ſteady in our Belief of the thing. The firſt of them occurs, *Exod.* 34. in the Perſon of renowned *Moſes.* His Face ſhined while he was upon Earth; and at the ſame time that it was corruptible Fleſh, did glitter with ſo bright a luſtre, that his fellow Mortals were

not

not able to behold it. And therefore whenever he talked with them, he was fain to wear a Vail, over the dazeling ſplendours of his radiant Countenance.

The other St. *Mat.* 17. in the moſt adorable JESUS, at his transfiguration. For then his bleſſed Face did not only ſhine, but ſhine as the Sun. And yet at the ſame time that it was ſo glorious and wonderful refulgent; it was but the Fleſh that it us'd to be. And therefore, ſays the Evangeliſt, μετεμορφώθη, he was changed in *form* only. And Saint *Luke* giving account of the ſame thing, does it in theſe words, ἐγένετο τὸ εἶδος τοῦ προσώπου αὐτοῦ ἕτερον, the faſhion *of his Conntenance was altered*; but nothing elſe. And if the Body of Man, in this ſtate of Mortality and Imperfection, could put on a ſhining, yea, a Sun-like Glory; by a change of its form or faſhion, its modes or qualities: then how eaſie will it be for a throughly changing modification, to ſuperinduce ſuch an alteration in all its Properties, at the Reſurrection; as is neceſſary to make it glorious in its full capacity. Though when all is done, it ſhall ſtill be but (* as the Father calls it) *caro angelificata, angelified fleſh.* Fleſh as pure, that is, and as ſpiritual, as thoſe Bodies which the Angels wear.

* *Tertul. lib. de Reſur.*

And truly if Chymiſts by the force of Fire, can reproduce Flowers (as we are told they can) out of their own Aſhes: and the Herbs that Beaſts have eaten, out of their Excrements, by the Architectonic parts latent in the ſame: then well may the Philoſopher argue and infer; what may not the laſt and dreadful ἐκπύρωσις, Burning of the World (which will eaſily reſolve things into their firſt Seminals) be able to do, towards the reſtoration of the dead Bodies of Men, or their happy refloreſcence out of their Duſt and Ruines? Eſpecially if he be a Chriſtian,

ſtian, and thinks the DEITY will ſtrike in with this melting Flame, and ſtrengthen and conduct it, in its mighty energy, to that very ſtrange and wonderful effect.

I ſay, if he conſiders, That the DEITY will ſtrike in, and ingage in the caſe. For I would be loth to have a ſiniſter miſconſtruction made, of what has been ſpoken. I mean, by being thought to impute the great and miraculous work of the Reſurrection, to the force of Nature only; or the perfect explication of the manner of its accompliſhment, to the light of Philoſophy. All that I have ſaid, amounts but to thus much, That Philoſophy helps us ſomewhat better to conceive of the fecibleneſs of the thing: as Nature helps much to illuſtrate its futurity, and alſo in ſome meaſure to exemplifie or repreſent it.

That Nature does thus, is clear, in the Judgment of *Minutius Felix*, an Author of equal elegance and ſolidity, in what he wrote; only pity it is that he wrote no more. His words, to this purpoſe, are theſe, * *See how entire nature for our comfort ſtudies* (to exhibit or typifie) *a Reſurrection to come. The Sun ſets, and riſes again; the Stars go down, and return again; the Flowers fade, and revive again; Shrubs, after the fall, have Leaves again; Seeds, unleſs they rot, ſpring not up again; ſo the Body in the Grave conceals a greenneſs, as Trees in the Winter, under counterfeit drineſs. Why art thou ſo haſty, as if it could revive and return, while rawtiſh Winter laſts? We muſt alſo wait for* the Spring time *of the Body.* Thus far he. And in the ſame way of ſetting forth the

* *Vide quàm in ſolatium noſtri, reſurrectionem futuram omnis natura meditetur. Sol demergit, & naſcitur; Aſtra labuntur, & redeunt. Flores occidunt, & reviviſcunt; poſt ſenium, Arbuſta frondeſcunt; Semina, non niſi corrupta, reviveſcunt: Ita Corpus in ſeculo* [Sepulchro] *ut arbores in hyberno, occultant virorem ariditate mentitâ. Quid feſtinas, ut crudâ adhuc hyeme reviviſcat & redeat? expectandam nobis etiam corporis ver eſt.* In Octav. *pag.* 113.

the Reſurrection, by ordinary Revolutions and Renovations in Nature, do the pious and learned Fathers go: As *Tertullian*, *Epiphanius*, *Ruffinus*, St. *Chryſoſtom*, St. *Ambroſe*, St. *Auſtin*, *Theodoret*, *Damaſcen*, and others.

Now juſt as Nature ſhews the futurity of the Reſurrection, while it prettily adumbrates and prefigures it to us, by various and lively Symbols and Reſemblances; ſo Philoſophy, another way, leads us into quick and clear Apprehenſions of its poſſibility: I mean, while it makes it evident, to ſuch as conſider, that a Body diſſolved, muſt ſtill exiſt in the *Minutes* or little Particles of it. And that the coarſeſt matter (even that of a cadaverous Body it ſelf) by a meer *Phyſical* change of its Modes or Qualities, may be made into a Subſtance moſt fine and glorious, and yet be really and eſſentially the very ſame it was.

But ſtill I ſay, as before, that this mighty Work, as to the main of it, is to be done by the Hand of GOD. And if any will maintain, That it cannot be; it behoves them, as the Chriſtian (*a*) Philoſopher has ſaid, to ſhew that it is, ἢ ἀδύνατον τῷ Θεῷ, ἢ ἀβούλητον, either above GOD's Power, or againſt his Pleaſure, πάλιν ἑνῶσαι καὶ συναγαγεῖν, *to gather together* the fragments and ſcattered materials of dead Bodies, and to *reunite* them, πρὸς τὴν ἀνθρώπων σύστασιν, for the *Conſtitution* or Inſtauration *of Men*. To him it ſolely belongs, to give ἑκάστῳ τὸ ἴδιον σῶμα, *to every one his own Body*, 1 Cor. 15. 38. And without his ſpecial directing and diſtinguiſhing Providence, impoſſible it muſt be, that particular Souls ſhould all recover the proper Ingredients of their reſpective Bodies, out of that unſpeakable blend and confuſion into which they will be run, before the end of the World.

(*a*) *Athenagoras, de Reſur. mort. p.* 175, 176. The whole Treatiſe is very well worth any Scholar's ſerious reading.

World. But ſince GOD will be pleas'd (to help Nature in the caſe) to put his Omnipotent Hand to the Work, we need not doubt but it ſhall be effected. *Why ſhould it be thought a thing incredible with you, that God ſhould raiſe the Dead?* ſaid St. *Paul*, *Act.* 26. 8. If any one elſe were to raiſe them, the doing of it might well be thought, ἄπιστον, *a thing incredible.* But ſince GOD will undertake it, why ſhould it be thought incredible? yea, even, παρ' ὑμῖν, *with you*; that is, King *Agrippa*, and *Feſtus*; to whom the Apoſtle was making his Defence, in open Court, at *Cæſarea.* Nor truly will the Queſtion ſeem improper, albeit they were Heathens. For though the Expreſſion, above remembred, be true; *the Reſurrection of the dead, is the confidence of Chriſtians*; none ever believed it ſo fully and firmly as they: yet that the very *Ethnicks* had a dim knowledge of it; and yielded a faint kind of aſſent or credit to it; we have grounds upon which to conclude. For

(a) *In Ancor.*

First, (*a*) *Epiphanius* tells us, That certain Nations had a cuſtom, to bring Meat and Drink to the Graves of the Dead, and in token that they expected they ſhould one day riſe, to invite them to it thus, Ἀνάστα ὁ δεῖνα, φάγε καὶ πίε, καὶ εὐφράνθητι. *Ho, riſe again, eat, and drink and be glad.*

And then *Athenagoras* informs us, That it was not only the perſuaſion of Chriſtians, that the Dead ſhall riſe again, but alſo of (*b*) many Philoſophers. And if they believed it, no wonder that Nations ſhould do the ſame; as being apt to be led by them, and to receive their Instructions: As they again ſurely would not fail to recommend a Doctrine of ſo high a nature, and excellent a uſe unto thoſe about them. Let me mention but a few of thoſe Philoſophers,

(b) Οὐ καθ' ἡμᾶς μόνον ἀναστήσεται τὰ σώματα, ἀλλὰ καὶ κατὰ πολλοὺς τῶν φιλοσόφων, *Legat. pro Chriſt. ſub fine.*

phers, whoſe commendation it is, that they inclined to a belief of the Propoſition, That the Dead ſhall riſe at laſt.

Democritus was of this mind, as we learn from *Pliny*'s checking him for it (*Nat. Hiſt. lib. 7. cap. 55.*) Where he calls *Democritus*'s Opinion of the Body's riſing again, a *Vanity*; and argues not only againſt that, but alſo againſt the Soul's exiſting in a ſtate of Separation.

Zoroaſter is ſaid to have taught, That there ſhall be a general Reſurrection of the Dead. And *Clemens* would make him a kind of *Earneſt* of it. For where (*c*) *Plato* relates of one *Herus Armenius*, That he was (*c*) *Lib. 10. de Repub.* ſlain in War, and lay ten days among the other dead Bodies; and being taken away, and carried home to be buried, on the twelfth day, when he was laid upon the Funeral Pile, he roſe to life again: (*d*) *Clemens* will (*d*) *Strom. l. 5. p. 599.* have this *Herus* to be *Zoroaſter*.

Plato, in his *Phædo*, concerning departed Souls, pronounceth, πάλιν γε δεῦρο ἀφικνῦνται, καὶ γίγνονται ἐκ τῶν τεθνεώτων, *They come higher again, and are made alive from the dead.* And the ſame thing he affirms in the very next words he writes, πάλιν γίγνεσθαι ἐκ τῶν ἀποθανόντων τοὺς ζῶντας. And a little after he inforceth what he had ſaid, by a more confident Aſſertion yet; (*e*) *Theſe things we acknowledge, not becauſe we are deceived, but becauſe there is indeed a returning to life, and a reviving from the dead.* Whereupon *Marſilius Ficinus* declares, *Videtur mortuorum reſurrectionem vaticinari. He ſeems to foretel the Reſurrection of the dead.* And by and by adds, *Tandem ingenii quadam fiduciâ reſurrectionem aſſerit. At length he aſſerts the Reſurrection with a kind of ingenious boldneſs.*

(*e*) Ἡμεῖς τὰ αὐτὰ ταῦτα οὐκ ἐξαπατώμενοι ὁμολογοῦμεν, ἀλλ' ἔστι καὶ τῷ ὄντι καὶ τὸ ἀναβιώσκεσθαι, καὶ ἐκ τῶν τεθνεώτων τοὺς ζῶντας γίγνεσθαι.

D 2 *Seneca*

Seneca does not tell *Lucilius* in plain terms, That the Dead shall rise: Yet he does very little less, when he writes to him in these words; *(f) Death, which we are so much afraid of, and so loth to submit to, does but intercept life, not take it way. The day will come, that shall bring us to light again. —— He that shall return* to the Body, *ought to go out of it with an even mind.*

(f) *Mors, quam pertimescimus & recusamus, intermittit vitam, non eripit. Veniet iterum qui nos in lucem reponet dies. —— Æquo animo debet rediturus exire. Ep.* 36.

The *Stoicks* (how rudely soever some of them disputed against the Resurrection, when St. *Paul* preached it at *Athens*, and huff'd and scorn'd both the Teacher and his Doctrine) did ever implicitly assent to and maintain it. For they held, That the Fire of the General Conflagration, which is to happen, κατ' περίοδον, *at the time appointed*; will be, πῦρ καθάρσιον, *a lustrative or purifying Fire*: and that the Effect or Consequent of the Purgation it shall make, will be, τῦ παντὸς διακόσμησις, *a new modelling of the World,* a restitution of the Universe to its pristine order, and of all Persons to their *quondam* conditions. So that herein, as *Origen* truly notes, *(g) Though they use not the Name of the Resurrection, they declare the Thing,* and allow it.

(g) Κἂν μὴ ὀνομάζωσιν τὸ τῆς ἀναστάσεως ὄνομα, τὸ πρᾶγμά γε δηλοῦσι. *Cont. Cels. lib.* 5.

Nor do the *Peripateticks*, in this matter, come behind those of the *Porch*, if we may judge of the Sect, by one of their School; I mean *Theopompus*. Concerning whom, *(h) Diogenes Laertius* has recorded, That he said, καὶ ἀναβιώσεσθαι τὺς ἀνθρώπους, καὶ ἔσεσθαι ἀθανάτους, *That Men* shall *return to life, and be immortal.* And he farther gives us to understand that one *Eudemus Rhodius* was of the same mind. And that *Theopompus* his Opinion in this, was, κατὰ τὺς Μάγυς, agreeable to the

(h) *In Proœm. pag.* 3.

the Wiſe Men. And that any of the Sect we are now ſpeaking of, ſhould aſſert the Reſurrection, may ſeem leſs ſtrange, if *Clemens* of *Alexandria* his report be true; who tells the World, (*i*) *That the* Peripatetick *Philoſophy depends upon the Law of* Moſes, *and the reſt of the Prophets.*

(i) Τὴν Περιπατητικὴν φιλοσοφίαν, ἐκ τῆς κατὰ Μωσέα νόμου, καὶ τῶν ἄλλων ἠρτῆσθαι Προφητῶν, *Strom. lib* 5. *pag.* 595.

Now when Philoſophy bids thus very fair for the Body's Reſurrection, is it not great pity that famous Divines ſhould perplex the Article, by inſiſting upon Niceties? Yet ſo ſome may ſeem to do (I mention not their Names in reverence to them) who will have the *Cuttings of the Hair*, and the *Parings of the Nails*, &c. to return to thoſe Bodies (when they ſhall riſe) from which they were once ſeparated. But one would be ready to think, that ſo great abundance of excrementitious Parts would be enough to make even a glorified Body deformed.

It cannot be deny'd indeed, but that the Apoſtle, diſcourſing of the Reſurrection of the Body, does poſitively affirm, δεῖ τὸ φθαρτὸν τοῦτο, **this** *corruptible muſt put on incorruption*; καὶ τὸ θνητὸν τοῦτο, *and* **this** *mortal muſt put on immortality*, 1 Cor. 15. 53. But *that* does not argue, that every *little thing*, which at any time was an *appurtenance* of the Body, is to be reſtored to it when it riſeth. If the riſing Body does but *wholly* ſpring up out of conſiderable parts of the *former* Body (out of ſuch parts as perhaps for their *drineſs*, *ſolidity*, &c. could never go to compound any other humane or carnous Body) this may come up ſufficiently to the Apoſtle's meaning. For ſo the raiſed Body may fairly be ſaid, to be *this mortal*, cloathed with immortality; and *this corruptible*, indued with incorruption; becauſe it will be the ſame with that

Body which was put off at death. The same, I mean, in a lax and Evangelical sense; and as much the same (I may venture to say) as our GREAT REDEEMER's was. For truly concerning *His* Holy Body, who can make it appear, that some part (and a good part) of its most precious Blood, was not left on the Whips, and on the Pavement; on the Nails, and on the Cross; on the Soldier's Spear, and especially on *Calvary*: after his most Sacred MAJESTY was risen? And yet if it were so, would it not strongly evince, that the Body raised, need not be the *same* with the Body dissolved, in a strict and nice, but only in a moderate or looser acception of Identity? Which if it might be allowed, whether it would not rescue the Doctrine of the Resurrection from some seemingly harsh intanglements and amusing difficulties, I leave to better Judgments to determin, being resolved my self not to be definitive.

The Schools (as is well known) are very strict here; contending that the same *Numerical* Body, in the rigidest sense, shall rise again. That the Hairs, Nails, Blood and Humours, must all return to the Body, and rise with it. But then let it be noted, in check to this Opinion, that it can ill be reconcil'd to another common one of theirs: That is to say, that the Bodies of Saints shall all arise, (*a*) *in juvenili ætate*, as *in youthful age*, and be *ejusdem statura, of the same stature* or proportion. So that Infants and Children shall have their Bodies as large, as if they had died, *in termino augmenti* (to use their word) in the *acme* or *full pitch of their perfect Growth*. Now if the self same, or very *Numerical* Body, shall rise at last; how is it possible, that Bodies so different in Dimensions and Quantity at their Dissolution, should ever put on an equality of Stature at their Resurrection? To affirm it,

(*a*) *Aquin. Sum. Theol. Sup. Quæst.* 81. *Ar.* 1, 2.

it, is as much as to ſay, that the raiſed Body ſhall be exactly the ſame for Bulk and Quantity that it was; and yet be bigger than it was, as to Size; and have more in it than it had, as to Subſtance. Elſe how can Infant Bodies be improv'd into a parity with thoſe of adult Perſons?

The Orthodox *Jews* did of old believe there ſhould be a Reſurrection. Though ſome Heretical ones denied and rejected it. And others were ſhort and erroneous in the Faith of it, as not extending it to the latitude we do, while they appropriated it to the *Juſt*. And others again were phanſiful or whimſical about it; poſſeſs'd with that vain and odd conceit of גלגול מחלות *the volutation of Caverns.* (Whereby they fondly imagin'd, that thoſe of them who were buried out of the Land of *Iſrael*, ſhall be rolled through ſubterraneous hollowneſſes, to riſe in *Canaan*, or elſe could not riſe at all.) Though ſome, yea, many of them might be thus carried away; yet the ſound and orthodox *Jews*, I ſay, did ſteadily believe a general Reſurrection of the Dead. And therefore in ſerious perſuaſion and expectancy of the thing, they us'd to call the Grave, בית חיים *the houſe of the living*. To which (by the way) our LORD might allude, when He would prove the Reſurrection againſt the *Sadducees*, by that Paſſage out of *Moſes*'s Writings (which they allowed, though they caſt off the reſt of the Old Teſtament) *I am the GOD of Abraham, and the GOD of Iſaac, and the GOD of Jacob; GOD is not the GOD of the dead,* ἀλλὰ ζώντων, *but of the* **Living**. Where he could not mean only, that *the Souls* of thoſe Patriarchs were then alive (for that would have been no juſt Proof of ἀνάστασις νεκρῶν, *the reſurrection of the* **Dead**.) But he meant, that the Bodies of thoſe famous Men (conſiderable pieces of the Individuals) without which

they

they could never recover the intire Personality they once had, were to be reckoned among the חיים, or *living*; as being potentially alive, alive in GOD's Purpose or Decree; and most certainly to be raised up, being actually revived at last.

But that which I mentioned the *Jews* Belief of the Resurrection for chiefly, is this: It was an Opinion among them, That the Body which riseth at the Resurrection, shall commence and spring up from a Bone (in the Back) called לוז. Which Bone the *Rabbies* will have to be incorruptible; and alledge the *Psalmist*'s words for it; *He keepeth all his Bones, Unum ex iis non conteretur,* One *of them shall not be broken*, Psal. 34. 20. Now though this be nothing but an empty conceit, yet it gives us the measure of the *Jews* Thoughts in the case; and plainly hints, That they were not curious, or much concern'd about the same *Numerical* Body's rising again; but deemed it sufficient to have the rising Body, made out of that Body which is laid down at death, or out of any part of it.

9. Another Article of extraordinary consequence, is, *The Conflagration of the Earth.* Of this the Gospel speaks expresly, *The Earth and the Works that are therein shall be burnt up*, 2 *Pet.* 3. 10. A positive Prediction, say the Atheistical and Incredulous: But how shall it be verified? Where is there Fire enough to do it? Here Philosophy assists again, by giving in most satisfactory Information. For it well assures us, That there are vast Stores of Fire *about* the Earth, as well as Treasuries of Fire *in* it: Yea, that the little Pores of almost every Body, have abundance of this Element lurking in them. So that we may wonder that the fiery Principle does not forthwith break out, and set all on a flame; rather than that it should take place

place at laſt, and ſeizing upon the World (I mean this Terreſtrial one) burn it up by a furious and inextinguiſhable combuſtion; *it being* (κατακλυζόμενος πυρὶ, as *Epiphanius* ſpeaks it) *delug'd with Fire.*

10. In the Fifth place: The Goſpel gives account of *a fiery Puniſhment prepared for Reprobates*, and of the *Eternity* of that Fire which is to plague and torment them. So we read in the Final Sentence, *Depart from me ye curſed into everlaſting fire, prepared*, &c. An impoſſible thing, ſays the prophane Objector. No Fire can burn for ever; for where ſhould be a ſupply of Fuel to continue it? Indeed a great Pile may make a great Fire; and durable Fuel may make a laſting one. But that any Fire ſhould be *everlaſting*, is not to be thought; it would want fit *pabulum*, or matter to cheriſh it, and inable it to ſubſiſt. Were all Combuſtibles amaſſed together, and made into one heap; were the whole Earth (as large as it is) turned into a *fomes* or aliment for it, it could not hold always. No, it muſt ſpend apace, and waſte and burn out into Cinders or Aſhes; and ſo the Fire that dwelt in it, and preyed upon it, muſt be ſtarved and die. This is a Knot that Philoſophy unties with the greateſt eaſe; and not only allows what our SAVIOUR has ſaid, to be probable and true; but adds confirmation to what he has authenticated, by caſting in the Overplus of its own authority. For it tells us there is a certain matter in the World, that burns of it ſelf, and will burn for ever. It is a Natural and Eſſential Flame; Independent and Vital (I may ſay) as being able to ſupport and maintain it ſelf alive, without any kind of Fuel to preſerve and feed it: So fixt and permanent, that where it is got together in any good quantities, it may very well challenge

the Epithet of *Unquenchable*; which is at once the dreadful Title and Property, whereby the Gospel describes that Infernal Fire that is to be the Instrument of the Damneds torture. Which must needs have strange force and vehemence in it to excruciate them; as being of a most subtil and active, and so of a most piercing and raging Nature.

11. Once more let us reflect upon the *Usefulness* of Philosophy. It is very great *in reference to Divinity*; that is, as it attends it in an humble subordination and service of its Interests. Chiefly in helping to clear up its Difficulties, and to keep Absurdities from mingling with its Doctrines. For thô (as the learned Father truly says) (a)*The Doctrine of our SAVIOUR* (which is the Marrow of all Divinity) *be perfect, and stands in need of nothing* (to improve it) *as being the Power and Wisdom of GOD: Yet Philosophy coming over to it, though it makes not the Truth the stronger, it makes sophistical Argumentation against it weak; and by driving away deceitful wiles against Truth, may be called a necessary Fence or Fortification of the Vineyard.* But not to be tedious:

(a) Αὐτοτελὴς μὲν καὶ ἀπροσδεὴς ἡ κατὰ τὸν Σωτῆρα διδασκαλία, δύναμις οὖσα καὶ σοφία Θεοῦ: προσιοῦσα δὲ φιλοσοφία, οὐ δυνατωτέραν ποιεῖ τὴν ἀλήθειαν, ἀλλ' ἀδύνατον παρέχουσα τὴν κατ' αὐτῆς σοφιστικὴν ἐπιχείρησιν, καὶ διακρουομένη τὰς δολερὰς κατὰ τῆς ἀληθείας ἐπιβουλὰς, φραγμὸς οἰκεῖος εἴρηται καὶ θριγκὸς τοῦ ἀμπελῶνος. *Clem. Alex. Strom. lib.* 1.

Philosophy, we know, has ever been reputed the Handmaid of Divinity; and its Honour it is, as well as its Excellency, that it really is so. And indeed where true and substantial Philosophy does interest it self in the Affairs of Theology, and serve it with a decent and respectful Ministry, it becomes of singular advantage to it. For the most perplext and intricate Problems it has (provided they be explicable) are no where made so clear and intelligible, as where Philosophy

ſophy is permitted or employed to intepret them, or call'd in to aſſiſt in the Expoſition of them. Whereas on the contrary, where that is ſhut out, there is too much cloudineſs and darkneſs within. Even familiar things are made confus'd and abſtruſe, and wrapt up in a nightſom and remedileſs obſcurity.

This is obvious to any ſlight notice. For in thoſe Theological Syſtems or Tractates, where Philoſophy has no place (as the *Mahometan*, for inſtance) People are miſerably gull'd, and ſhamefully impoſed upon, by monſtrous Aſſertions and that in common and eaſie matters. Yea, in many things, the groſſeſt Extravagances paſs with them for ſublime and ſolid Notions; and the fulſomeſt Non ſenſe, for venerable Myſteries; they wanting wherewith to diſtinguiſh betwixt them. And all becauſe they are ἀφιλόσοφοι, deſtitute of Philoſophy; which would afford them ſuch κριτήρια, Rules of Judgment, as would inable them to diſcern what is true, and what is falſe; and to take exact meaſures of a juſt diſcrimination.

That the Virgin *MARY* conceived by ſmelling to a Roſe, and was afterwards delivered of Chriſt at her Breaſt. That the Moon once ſlipt into *Mahomet*'s Sleeve. That in the End of the World all things ſhall die at the winding of an Horn, even Angels themſelves. That the Accurſed then following *Cain* as a Leader, ſhall carry their Sins in Satchels at their Backs; the weight of which breaking down the *Bridge of Juſtice* as they go over it, the Delinquents, for their Puniſhment, ſhall fall into a River of Fire that runs under it. What prodigiouſly dull and heavy Figments are theſe? Yet by the zealous Proſelytes of the Epileptick Impoſtor, they are embraced not only as Rational and Conſiſtent, but even as Divine and Miraculous Truths. But were the light of Philoſophy let into the

 Alcoran,

Alcoran, it would ſoon chaſe away theſe, and all ſuch egregious and nauſeous Fooleries. And together with them, it would ſend blindneſs and ignorance packing too (which are their Parents and their Nurſes) making them flee before it, as Chaff before the Wind, or as the Shades of the Night before the Morning Sun.

And here if it might not be thought too unſeemly a Tranſition, to paſs from the *Turkiſh* to the *Chriſtian* Divinity; it would not be impertinent to remark *its* defectiveneſs in Philoſophy likewiſe in one particular. How much better might, *Hoc eſt Corpus meum*, *This is my Body*, have been underſtood; both as to the credit of the Propoſition, and its influence upon Chriſtians, if Philoſophy had paraphras'd or expounded it? For then the true Notion of an Humane Body, ſhould have ſhut out that numerous train of Abſurdities (beſides many, and great, and intolerable Inconveniences) which have followed the literal acceptation of the words.

12. Theſe are a few ſlight touches upon the USEFULNESS of Philoſophy. They might have been greatly both multiplied and enlarged. But being intent upon Brevity, I ſay no more of that nature. Only as I go along, give me leave to drop this Note by the way; What an auſpicious Providence is it, that has made ſome Perſons for ſeveral years paſt, ſo indefatigably ſtudious in Philoſophy, and has crown'd their Studies with ſuch happy ſucceſs! It looks as if GOD had ſtrange kindneſs for us, and deſigned to enrich and honour this Age with a very choice Bleſſing; even with a clearer underſtanding, not only of *Nature*, but in *Divinity* it ſelf, than the World has yet had.

Nor

Nor need we wonder that Philoſophy ſhould advance and flouriſh even among our ſelves; conſidering what encouragement it has lately met with. For whereas it is recorded by (a) *Ælian*, as a ſignal Reſpect to Philoſophers, That *Cæſar* us'd to call at *Ariſto*'s Door; and *Pompey*, at *Cratippus*'s: we have ſeen a Renowned MONARCH of our own, ſhew *higher* Favour, and give *better* Countenance, to that honourable and uſeful Sort of Men. For beſides that he erected them into a famous and well-contriv'd *Society*; that they might be a *ROYAL* one too, he was pleaſed to make HIMSELF Their PATRON and Their HEAD.

(a) Οὐκ ἀπηξίου Καῖσαρ ἐπὶ τὰς Ἀριστίωνος θύρας φοιτᾷν. Πομπήϊος δὲ ἐπὶ τὰς Κρατίππου. *Var. Hiſtor. lib.* 7. *cap.* 21.

13. Now we being in theſe hopeful thriving Circumſtances, it behoves us to be careful, that it may not be with us in this Caſe, as it frequently and unfortunately falls out in others, *That we Abuſe not Philoſophy*, which grows ſo faſt and high in noble Improvements. For ſuch is our looſe and unruly temper, that from fulneſs we commonly run into wantonneſs, turning our rich Plenty into Riot and Exceſs. It will be wiſdom to reſtrain or ſuppreſs this humour, and to keep our ſelves free from its evil predominance: elſe our very Remedy may become our Diſeaſe; and we may ſurfeit and be ſick of that wholſom medicine which the Ἀρχίατρος, or great Phyſician of our Minds, has given us to help their pitiable Infirmities.

14. And upon this account of *Abuſing* Philoſophy, they are mightily to blame that *Slight* it: I mean, *by ſpeaking or thinking little and unhandſom things concerning it.* Too many there are of the *firſt* Stamp, that *Speak unbecomingly* of Philoſophy. They cry it down

down as needless, and condemn it as superfluous; and tell us, there's no cause to trouble our selves about it, we having an higher and better Doctrine among us, which came down from Heaven. These are crooked Reflections, and garishly made, upon no grounds, and without reason. The language (at best) but of Prejudice or Ignorance, Mistake or Inadvertency. For though the Word from Heaven be infinitely preferible to Humane Learning; yet the one is most necessary in conjunction with the other: Else we may believe the Wise GOD would not have given us both: But since he has done so, it concerns us to take heed of slighting either. To reject Philosophy upon the account of Scripture, is just as discreet, as to resolve all Industry into Providence; and to take no thought in the least for our selves, because GOD careth for us. Just as rational as it would be for a Man to throw away all Candles, because there is a Sun; or to put one of his Eyes out, because he has two.

Clemens of *Alexandria*, upon strict Examination, will scarce be found clear of this Oversight. Not that he was for banishing or casting out Philosophy (he used as much of it as any Father) but he spake disparaging words concerning it: he diminisht it greatly, and detracted from it, by making it of a Base and Diabolical Extraction. For he fetches its Origin directly from *Dæmons*, and talks as if lewd and lascivious Spirits were the Authors of it; while they imparted it to Women, and these to Men (as *Sampson*'s Paramour did his Riddle to the *Philistines*.) Thus says he in one place; (a) *Philosophy was not sent from the LORD; but came into the World, stoln and delivered*

(a) φιλοσοφία οὐκ ἀπεστάλη ὑπὸ Κυρίου. ἀλλ' ἦλθε κλαπεῖσα, ἢ παρὰ κλέπτου δοθεῖσα. εἴτ' οὖν δύναμις, ἢ ἄγγελος μαθών τι τῆς ἀληθείας καὶ μὴ ἐμμείνας ἐν αὐτῇ ταῦτα ἐνέπνευσεν, &c. *Strom. lib* I.

by

by a Thief. Whether therefore some Power or Angel having learned somewhat of the truth, and abiding not in it, inspired these things, &c. And in another place; (*a*) *The Angels, whose lot was above, being fallen down into Pleasures, decl.red Secrets, and such things as they knew, to Women; while other Angels would rather have conceal'd them, and reserv'd them till the coming of the* LORD. *And from thence did the Doctrine of Providence flow, together with the discovery of high things.* I would fain believe that he relates the fanciful conceits of others, rather than his own Opinion or Judgment. And I meet with a certain Passage, in the (*b*) First Book of his *Stromata*, that hints as much. For speaking there of the *Greek* Philosophy, and saying, *All Philosophy was inspired by certain inferiour Powers*, he brings it in with an ἔνιοι ὑπειλήφασιν, *some have thought so.* And saying, that ἐκ τοῦ Διαβόλου τὴν κίνησιν ἴσχει, *it was set a foot by the Devil*; he makes way for it, with ὡς ἄλλοι βούλονται, *as others will have it.* But whether this learned Person spake his own sense, or other Mens, or both at once, we have another as learned as himself, who pronounceth much better, and more truly in the case; namely, (*c*) *Whatever Philosophers have spoken well, GOD manifested to them.* By the light and use of their Reason; that is, not by immediate supernatural Inspiration: It was the product of Study, not of Revelation. According to which high Derivation of Philosophy from GOD himself, we are obliged at least to *speak* fairly and honourably of it.

(*a*) Οἱ ἄγγελοι οἱ τὸν ἄνω κλῆρον εἰληχότες, κατολισθήσαντες εἰς ἡδονὰς, ἐξεῖπον τὰ ἀπόρρητα ταῖς γυναιξὶν, ὅσα τε εἰς γνῶσιν αὐτῶν ἀφῖκτο, κρυπτόντων τῶν ἄλλων ἀγγέλων, μᾶλλον δὲ τηρούντων εἰς τὴν τοῦ Κυρίου παρουσίαν. ἐκεῖθεν ἡ τῆς προνοίας διδασκαλία ἐῤῥύη, καὶ ἡ τῶν μετεώρων ἀποκάλυψις. *Strom. lib.* 5.

(*b*) *Pag.* 309.

(*c*) Ὁ Θεὸς γὰρ αὐτοῖς ταῦτα, καὶ ὅσα καλῶς λέλεκται, ἐφανέρωσε. *Orig. cont. Celsum, lib.* 6.

Which

Which that we may do, we must so inwardly esteem and value it, as by no means to *slight it in our thoughts*. For mean and unworthy thoughts will soon break out into low and disparaging Speeches. Let none therefore depretiate or vilifie Philosophy, so much as in *thought*. He that does so, deserves to be branded for a Wretch. And accordingly indeed he is stigmatized by (d) *Plato*: markt out for a pitiful sneaking Soul, that is sunk below the Dignity of his raised *Species*; and in token of his degeneracy should write, MAN, no longer. And most justly. For he that is so silly as to despise Philosophy, does thereby condemn the improvement of Reason; and consequently sets his mind against Reason it self, the specifick perfection, as well as the Crown and Glory of Humanity. And why should *he* wear the Title of that Nature for an Ornament, who is an Enemy to its excellency?

(d) Ὁπότε γάρ τοι τὸ φιλοσοφεῖν αἰσχρὸν ἡγησαίμην εἶναι, οὐδ' ἂν ἄνθρωπον νομίσαιμι ἐμαυτὸν εἶναι. *In Amator.*

15. But besides such *mental* and *verbal* wrongs that are done to Philosophy, which amount to a considerable undervaluing of it: there are also real and *actual abuses* put upon it, to the unreasonable prejudice of our selves, and that, while we make it minister to vanity and extravagance. And this is done *Two* ways chiefly.

The *First* is, When we *set it too low*, and depress it beneath its generous Worth. When we inslave it to Phansies and ridiculous Whimsies. When we put it upon the Investigation of things that *are not*; or if they *be*, are but Smoke and Shadows; not worth the thinking of, much less the searching after. And here the *Schools* are more than a little to blame. For

For they ſometimes contend *de lanâ Caprinâ*, about great and loud nothings: and debate and wrangle with heat and fierceneſs, upon the moſt empty and frothy Queſtions. Thus (to inſtance in the matter of Angels) they diſpute whether an Angel can be moved *locally* or no. And yet at the ſame time, their Philoſophy will not admit an Angel *to be in a place*. And yet they ſtill go on to argue, whether it be moved *from place to place, pertranſeundo media, by paſſing through the intermediate places*. And then they farther Query, Whether (if an Angel moves through intermediate places) it be done, *in tempore, vel in inſtanti, in time, or in an inſtant:* and ſo whether he cannot as ſoon move from one Pole of the World to the other; as croſs a Room not ten Foot wide. This is to trifle with Philoſophy, to put tricks upon it, or to tantalize it. To ſtarve it under the pretence of treating it: or to make it a *Chamelion*, and to feed it with Air. It is perfectly to ridicule its Gravity: and to turn its Glory into ſcorn and contempt. To pull it down from its glittering and ſtately Throne of Reaſon, and to place it amongſt the Vaſſals or Lacqueys to humor; by tying it to wait upon empty impertinencies, with uneaſie ſervility, People that do thus affront and abuſe Philoſophy, by ingaging it (as I may ſay) in lamentable druogery; in ſpinning out thin and Cobweb notions, in cutting up lean and aiery niceties, and in ſearching into ſuch jejune and ſorry Queſtions, as afford little or nothing worth its regard; do thereby leſſen and degrade *themſelves*, as well as *that*. And that ſo far, that I know not how to call them, nor what ſufficient names of Diminution to beſtow upon them. They are Sons of Sophiſtry, Pragmatical Speculators, Promoters of

F Subtilties,

Subtilties, Metaphysical Brokers, Dissectors of Ideas, Factors for Levity, and Problematical Coiners. They are Fanaticks in Philosophy, and Conspirators against it, and Persecutors of it, and Worms and Moths to it; and if they chance but to get into its rich Wardrobe, are for eating and spoiling all that is good in it. They are Men that seek for Knots in Bulrushes; and if they cannot find them there, will be sure to make them. That will beat solid Learning, as thin as a *Superficies*; mince it as small as Mathematical Points; grind it into finest Powder of Atomes; turn it first into shadows, and then into nothing. In a word, whatever they seem to *be*, they are but meer Skeletons or Ghosts of Philosophers; which commonly haunt wise Notions, till they fright them out of their Wits. And whatever they seem to *do*, they make but a noise and a great pother; and in the dust that they raise, many times lose what they pursue. So that at best they are but Cyphers, or insignificant Persons; idle when they are most active; and useless when they are most studious; and that I may borrow their full character from * *Plato* in short, φλυαρῶν φιλοσοφοῦντες, *they* only *tittle tattle in Philosophizing*.

* *In Amator.*

16. The *Second* way of *abusing* Philosophy *actually*, is by forcing it upon the contrary extreme. By straining it too much, and stretching it too far, and *screwing it too high*. By ascribing such a mighty influence to it, as (is conceited) may help Nature over all kind of Difficulties: or put her into a way of solving every *Phænomenon* almost, by her own intrinsick force and vertue, by her own established Laws and Conducts; without recourse to any other Principle or Oeconomy. As if because it

does

does *many* things, it muſt therefore do *all*; and nothing could be wrought without its efficiency. By which means, it is not only arm'd with an energy, far above what it can rightfully claim; and preſumed efficacious beyond all its proper Ability can extend to: but moreover is ſet up in oppoſition to Providence, and made to derogate from a Power, that is infinitely ſuperior to its own; by challenging to its ſelf, what that alone is able to effect. Or if it be not made to arrogate ſo much to it ſelf, as downright to claim the power of Miracles; yet then it is ſuppoſed to intermeddle too far in the working of them: and ſo to intrench upon Omnipotence, where it does not exclude it.

And therefore that cavilling or captious Objection againſt the Philoſopher, which *Plato* complains of, applied to him that runs Philoſophy too high, and attributes too great a virtue to it; may paſs for an allowable Cenſure, in ſome meaſure at leaſt. * *That he ſearches into things above and below, and minds not the GODS.* To which reproof too many are obnoxious (though it be frequently thrown out at random alſo, and in its wild projection ſometimes hits ſuch as do not deſerve it.) They are ſo intent upon inferior Cauſes, as to overlook the Supreme. And that where his hand is not only immediately ingaged, but alſo in a manner even viſibly operative, by way of Divine or ſupernatural agency.

* Ὅτι τὰ μετέωρ. κ̀ τὰ ὑπὸ γῆς ζητεῖ, κ̀ Θεὺς μὴ νομίζει. *Apol. Socr.*

But they that impute too much to Philoſophy, however they may think they befriend and honour it; they thereby do it very great diſſervice. For while they cry it up more than they ought; they make others ſlight it, and give it leſs than its due. A curious Statue that is delicately wrought; finely proportioned, that is, admirably featured, rarely and exquisitely graven or carved: How many does it take, and how mightily does it please them! But let this pretty Artificial Piece be converted to an Idol, and dress'd out, and set up to be worshipped; let more be said of it than is really true, and more ascribed to it than it possibly can do: and this turns the value they had for it before, into scorn and hatred; making nothing more vile and detestable than it. Even so the easiest and readiest way to depress Philosophy beneath its true worth, and to bring it down into disesteem, not to say, into lowest contempt with Men; is for them to idolize it (as it were) and adore it, by applauding it extravagantly, and elevating it beyond its proper merit and capacity. For so they who before had a meet respect and veneration for it, will quickly despise it, and even swell with antipathy and indignation against it. Let all that are too great admirers of it, consider if thus they be not enemies to it, and preter-intentional detractors from it.

Thus some Devotoes and overweening Magnifiers of it, do solve the overthrow of *Sodom* Philosophically. They will have nothing at all of miracle in it; but hold, it was consumed by Tempest. Accidental Lightnings, that is, from Thunder-clouds above, kindled subterraneous Fires about it; the ground whereon that City was built, being of a very bituminous substance. But they who shall peruse the sad Story of the Calamity, and well observe the special hand that GOD is noted to have in it: and also seriously consider the solemn Dialogue betwixt HIM and *Abraham* about it; and the wonderful deliverance of righteous *Lot* from it: must certainly be of another Judgment, than to think it proceeded from nothing but Lightnings in the Air, and

and Sulphureousness in the Earth; and that it was the simple effect of a meer natural Causality. And truly that one expression, * *The LORD* rained *upon* Sodom *Brimstone and Fire from the LORD out of Heaven*, does plainly intimate, that it was Miraculous. * Gen. 19. 24.

Others bear us in hand, that there was nothing extraordinary in the destruction of *Pharaoh* and his mighty Army: but that they were drowned by pure oversight, and the common course of those famed Waters. Entring the Bay, that is, when the Flood was withdrawn, and the Waters at an Ebb; they marched too far, and carelessly continued too long therein: even till the retired Sea, returning upon them in an impetuous Tide, swallowed up the King and his Host in an Instant. A likely supposition! when we are told expresly from the Mouth of GOD, *That he divided the Red Sea* לגזרים *into* Parts, or Segments, *Psal.* 136. 13. Where the *Jewish* Tradition is, That the Sea was divided into *Twelve* Cuts, according to the number of the Twelve Tribes; and so every Tribe passed through the Channel in a Lane by it self. But however that be nice and humorous (as is their other conceit, that seventy two Angels just, assisted in the Miracle; because the Nineteenth, Twentieth, and One and Twentieth Verses of the *Fourteenth* of *Exodus*, do each of them contain so many Letters) yet that the Red Sea was actually parted asunder, is clear from GOD's Command to *Moses*, at the *Sixteenth* Verse of that Chapter; *Lift thou up thy Rod, and stretch out thine hand over the Sea, and DIVIDE it.* And accordingly it is said at the *One and Twentieth* Verse, ויבקעו המים *the Waters were* DIVIDED. So divided, that as we read in the following Verse, they *were a Wall to them to them on their right hand, and on their left.* Which expression, to them that consider the situation of the Red Sea, and the course of journeying the *Israelites* were then in; does plainly discover, that the Waters at their passing through them, were perfectly divided. Else they could not have had them on both hands of them at once, at least not as a *Wall* to them. And that the Waters of *that* Sea, were as much divided as Waters could be; is farther manifest from what occurs in the Story of *Israels* passage over *Jordan.* Where it is said, That *the Waters* (of that River) *which came down from above, stood and rose up upon an heap: and those that came down towards the Sea of the Plain, failed, and were cut off,* Josh. 3. 16. So that if the Waters of the Red Sea were but served thus, they could not possibly be more really divided. And that they were thus served, is plain from the Testimony of the HOLY GHOST, *Josh.* 4. 23. *The LORD your GOD dried up the Waters of* Jordan *from before you, until ye were passed over, as the LORD your GOD did to the* Red Sea.

Nor is it improbable (by the way) that proud *Chencres* (that *Egyptian* Monarch, who at last was overwhelmed with the Waters of this Sea) might derive his obstinacy (the occasion of his ruine) from what we are now speaking of: I mean from his *setting Philosophy too high.* He might be strongly opinion'd, that *Moses*'s Works were no Wonders at all; no more than what Nature her self could do, if manag'd by Philosophers. And then *Jannes,* and *Jambres,* and the Magical Crew, (*Pharaoh*'s Philosophers) pretending imitation of the Man of GOD, by their Juggling Tricks, might confirm the Tyrant in his mistaken thoughts: and convince him that

that he made a right Judgment of things, when he believed them to be of an ordinary ſtrain, and placed them to the Accounts of Nature and Philoſophy. Or if *Moſes*, in ſome matters, out-did the Philoſophers of the haughty King; yet this he might conclude the reſult of his Breeding. For knowing he was brought up in his Royal Court, he could not but be ſenſible he had ſingular Advantages; and in caſe he improved them (as he had reaſon to think he did) might well outſtrip his notableſt Antagoniſts, upon the ſcore of his Education. And truly as *Philo* informs us, That he * *attain'd to the top of Philoſophy*: ſo † Scripture aſſures us, That he *was Learned in all the Wiſdom of the* Egyptians. Which *Wiſdom*, the ſame *Philo* gives a particular Account of, *in his life of* Moſes. And therefore *Pharaoh* might juſtly take him to be more skilful than any of *his* Philoſophers; and yet preſume, that the Works wherein he excell'd all his artful Men, were not owing to Divine Power; but to ſome knack he had of ordering Nature more dexterouſly, than they could do. And thus the prejudice once ſprung up, might be deeply rooted in the Prince's Mind. And therefore, though he *felt* as well as *ſaw*, the mighty Prodigies (for they were ſorely afflictive, as well as ſtupendious) yet he ſtouted it out a long time againſt them. Nor would he at laſt have buckled to releaſe the *Hebrews*, had not frightful Death ſhown it ſelf on the Tragical Stage; and alſo come up ſo near him, as to ſtrike his Firſt-born and Heir apparent to the Crown.

* Μωσῆς δὲ καὶ φιλοσοφίας ἐπ' αὐτὴν φθάσας ἀκρότητα. *Lib. de Mund. Opif.*

† Act. 7. 22.

And

And the ſame thing is atteſted by *Origen* of the *Egyptians* in general thus far, That they did not look upon *Moſes* as doing the ſtrange things he did, by power from above. (*a*) *Thô* (ſays he) *they do not abſolutely deny the Wonders wrought by* Moſes, *yet they declare they were done by deluſion, and not by Divine ſtrength.* And the *Sorcerers* or *Magi* taught them as much; ſays the learned *Jew* that wrote *Moſes* his Life. And therefore when his Rod was turned into a Serpent, and the Multitude of Spectators (among whom was the King himſelf and his Princes) were ſtruck at once with amazement and fear, inſomuch that they fled: the ſame Writer brings in the (*b*) *Sophiſters and Magicians*, ſpeaking thus, *Why are you affrighted? we are not ignorant of ſuch things, but uſe the like Art our ſelves in publick.* Intimating, they thought *Moſes* wrought the Miracle by pure Legerdemain, or Magical Craft, wherein they were verſed. Which is alſo farther inſinuated by that Proverbial Taunt, wherewith they flouted the renowned Heroe (as we find it in the *Talmud*) *Thou bringeſt Straw into* Afra: that being a place in *Egypt* where Straw abounded. Meaning, it was vain for him to play *Hocus-Pocus* Tricks, or to practiſe Inchantment, in a Land ſo ſtockt with the ſame already.

(*a*) Τὰς μὲν Μωσέως περιστὰς δυνάμεις ἐ παντελῶς ἀρνέμενοι, φάσκοντες δ᾽ αὐτὰς γοητείᾳ καὶ μὴ θείᾳ δυνάμει γεγονέναι. *Cont. Celſ. Lib.* 3.

(*b*) Σοφισταὶ δ᾽ ὅσοι καὶ Μάγοι παρετύγχανον, τί καταπέπληχθε εἶπον; ἐδ᾽ ἡμεῖς τῶν τοιέτων ἀμελετήτως ἔχομεν, ἀλλὰ χρώμεθα τέχνῃ δημιεργῷ τῶν ὁμοίων. *Philo in vit. Moſ.*

Nor can I paſs by thoſe ſawcy Reflections, that lewd *Celſus* makes (in favour of Philoſophy) upon our SAVIOUR's Miracles; though it carries me a little farther ſtill out of my way, to take notice of his Baſeneſs. They were much ſuperiour to the Wonders

ders of *Moses*. Yet that rude *Epicurean* would fain argue them down into the hateful Rank of Prestigious Impostures, and make the HOLY JESUS no better than a Conjurer. Yea, having gotten the sacred Story by the end, of our LORD's Flight into *Egypt*, he perverts it most shamefully, to make it countenance that black and hellish Reproach which he would have fastned on his GLORIOUS MAJESTY. For he blasphemously affirms, That he was *(a) brought up in an obscure manner, and being Lett for a Servant thither, grew skilful in the strange Feats* of that Nation; and then *returning from thence, by the Feats he could do, gain'd himself the Name and Repute of a GOD.* And yet still (says the Wretch in another place) he was but a *(b) Juggler*, and as such a one, went up and down *dishonourably begging, and getting his livelihood* by what he could do by Sleight-of-hand. Now whither tends this? Why, as it is all but a Cast of *Celsus*'s profound Philosophy; so the drift of it was but to advance Philosophy, and set it too high: to exalt it, that is, above the Christian Doctrine, and to maintain it in way of Opposition to that.

And thus, (to come home to our purpose at last) Some have set Philosophy too high, in reference to the *Flood*: I mean that great and general Flood, which put a disastrous period to the First World. For they held it proceeded from Second Causes, in such a manner as reflects upon the First: in such a manner, that when, to do the greater honour to Philosophy, they attempt, by the help of it, to explain *how*; the Explication grievously impeaches Scripture, and charges it very boldly and unhandsomly; a thing by no means to be endured.

(a) Φησὶ γὰρ αὐτὸν σκότιον τραφέντα, μισθαρνήσαντα εἰς Αἴγυπτον, δυνάμεών τινων πειραθέντα, ἐκεῖθεν ἐπανελθεῖν, Θεὸν δι' ἐκείνας τὰς δυνάμεις ἑαυτὸν ἀναγορεύοντα. *Orig. cont. Cel. lib.* 1.

(b) Ὁ τοῦ Θεοῦ παῖς οὕτως ἀγείρεις ἀγυρτῶς. *id. ib.*

endured. For though Philosophy (as has been said) be eminently serviceable to Divinity; and that in its noblest and most important Articles: unless they be such as are absolute Mysteries (and so naturally as unintelligible to meer Reason, as finest Speculations are imperceptible to Sense) yet it must not be allow'd to clash or interfere with it in the least; especially in its holy Foundations or Principles, the Inspired Oracles. For so the Hand-maid would pertly usurp over her Mistress; and forgetting her duty, proudly domineer in her Station of Obedience.

17. And this is too much the Fault of *The Theory of the Earth*. It pends too hard against the Sacred Scriptures, and advances to an intrenchment upon Divine Revelation. Which will evidently appear, in several Particulars, in the Sequel of our Discourse.

18. It abounds with Philosophy indeed; and the Philosophy it contains is well delivered. But it is not justly regulated, and kept within due Limits. For it runs so fast, and is driven so far, that it treads unseemly, and unsufferably too, upon the heels of Truth; even of that most Divine and Infallible Truth which was spoken by GOD; and therefore to be infinitely reverenc'd of Men.

19. Now this Irregularity I apprehended so great, that the reverence I bear to that *Holy Volume*, whose Contents are no other than the Doctrines of Heaven, ingaged me in drawing up the ensuing *Exceptions*, and then in publishing them. Though I must own too, that I was much encouraged in the Undertaking, by the *Theorist*'s ingenuous and frank Invitation: (a) *Whosoever, by solid Reasons, will shew me in an Error,*

(a) Theory, pag. 288.

ror, and undeceive me, I ſhall be very much obliged to him. This I ſhall endeavour to do, with all Sincerity; and that only as a Friend and Servant to Truth. And therefore with ſuch Candour, Meekneſs and Modeſty, as becomes one who aſſumes and glories in ſo fair a Character: And alſo with ſuch Reſpect to the *Virtuoſoe* who wrote the *Theory*; as may teſtifie to the World, that I eſteem his Learning, while I queſtion his Opinion.

20. And that our Work may be done with the more eaſe and order, it ſhall be proſecuted in a Method cut out to our hands; and ſhaped according to that Recapitulation of the *Theory*, which we find ſet down in the *Second Book*, and the *Ninth Chapter*, in theſe words:

That there was a Primitive Earth, of another Form from the preſent, and inhabited by Mankind till the Deluge: That it had thoſe Properties and Conditions that we have aſcribed to it, namely, a perpetual Equinox and Spring, by reaſon of its right Situation to the Sun; was of an Oval Figure, and the exterior face of it ſmooth and uniform, without Mountains, or a Sea: That in this Earth ſtood Paradiſe, the Doctrine whereof cannot be underſtood, but upon Suppoſition of this Primitive Earth and its Properties. Then, That the Diſruption and Fall of the Earth into the Abyſs which lay under it, was that which made the Univerſal Deluge, and the Deſtruction of the Old World: And, That neither Noah's *Flood, nor the preſent Form of the Earth, can be explained in any other Method that is rational, nor by any other Cauſes that are intelligible. Theſe are the Vitals of the* Theory, *and the Primary Aſſertions, whereof I do freely profeſs my full Belief.*

Against these *Assertions*, my *Exceptions* shall be levelled; and in the same order in which they stand.

21. So much for the First Chapter; which may be reckoned as an *Introduction* to the following Discourse. Which if any shall look upon as a Collection of Notes somewhat confusedly put together, rather than a formal, well digested Treatise, they will entertain the best or truest *Idea* of it.

CHAP.

CHAP. II.

1. *The* Hypothesis *of the Earth's* Formation *stated.* 2. *The* first Exception *against it,* It would have taken up too much time. 3. *The* World *being made in* Six Days. 4. *How there might be* Light *and* Days, *before there was a* Sun. 5. *A* Proof *that the* Creation *was perfected in* Six Days *time.* 6. Numeral Cabbalism *cannot overthrow it.* 7. *The* Jews *in* Cabbalizing, *still allowed a* Literal meaning *to Scripture*; *only they superadded a* Mystical one, *never* contrary *to it.* 8. *Though were there a* Cabbala, destructive *to the* Letter *of* Moses's *Story of the Creation,* that *would not invalidate the Argument alledged.* 9. Moses's *Account of the Creation, runs not upon bare* Numbers, *but upon* Time. 10. *What Account the* Christian Church *has made of the* Cabbala. 11. *How it discovers its own* Vanity. 12. *The* Literal sense *to be kept to, in the Story of the Creation.* 13. *Where Scripture speaks so as not to be understood* Literally, *it is* sometimes *for* plainness sake.

1. AS every thing had a Beginning except One; I mean, that most perfect and glorious ESSENCE, who gave Being to all; so the Earth, among the rest, had its Origin likewise. This, none but Infidels, or Anti-Scripturists, can doubt; the Article being founded upon no less than Divine, which is the most firm and unquestionable Evidence. Could any Doubt of this Matter offer to form it self in our Minds, and to settle there, the very first Verse in the Holy Bible would not fail to drive it out from thence.

But

But then as to the *Way* of the Earth's Formation, we are more at a Loss, as being not so satisfactorily instructed concerning it. Here Providence seems to have left us to our selves; and for the improvement of both, remits us to the Conduct of Philosophick Learning, in some measure, and to our own Judgments. Only we must be careful that the *Idea's* we frame, be congruous to the Truths that came down from above; and are, or should be the Touchstone of all *Hypotheses* among Christians. Which, because the way of the Earth's Formation, according to the *Theory*, is not; it overthrows the first vital Assertion; which is this, *There was a Primitive Earth of another Form from the present, and inhabited by Mankind till the Deluge.* The latter Clause of it, touching the Earth's being inhabited till the Deluge, we do not question. The former part of it cannot stand, by reason the *Manner* of the Earth's Rise, which the *Theory* ascribes to it, overturns it. It is supposed to have proceeded thus.

The whole matter of the Earth and sublunary Heavens being confusedly blended together, in one fluid Mass or *Chaos*; the grosser and heavier parts thereof sunk down to the middle of it (as to the Centre of their Gravity) and constituted an interior Orb of Earth. The rest of the Mass about it, by the same Principle of Gravity, was divided into two Orders of Bodies; the one Liquid, and the other Volatile. The Volatile mounting above the Liquid, constituted the Air: The liquid mass swimming below (and incircling the inward Earth aforesaid) contained in it all Liquors originally belonging to the Earth. These terrestrial Liquors are of two kinds chiefly; either fat, oily, and light; or else lean, and more earthy. The lean and earthy Liquor made up the Element of common

mon Water: The Oily Liquor (which arose out of the Water as it purged it self) got above it, and floated upon it. And so the whole stood as in this Figure:

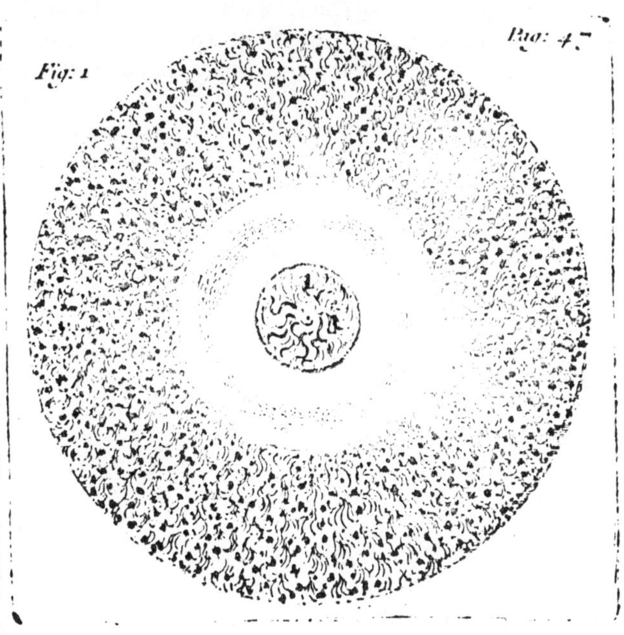

Where 1 denotes the fiery Centre of the Earth. 2 The Interior Orb of the Earth, composed of the grossest particles of the *Chaos*. 3 The Element of Water, or the Abyss. 4 The Oyly Liquor upon the surface of the Water. 5 The Mass or Body of the Air.

But

But this Body of the Air being at first very muddy and impure, through abundance of terrestrial Particles that, as fast as they could free themselves from the Air with which they were mingled, and in which they were intangled, they sunk downward: And meeting, in their descent, with the Oyly Liquor on the face of the Deep, there they stuck; and incorporating with that unctious Substance, made a certain Slime; or a fat, soft, light Earth, spread upon the Waters: Which growing thicker and thicker, by a continual accession of more terrestrial Particles, sliding down still out of the Air, as it purify'd it self; at last it came to its just Dimensions. And then waxing more dry and stiff, and firm and solid, in fine it attained to its due Consistency, and so became the First habitable Earth.

Thus have I briefly, but, I hope, truly represented the *Manner* of the Primitive Earth's Formation. If there be any thing of Mistake in the Description, it is altogether involuntary. But I think I have spoken the very mind of the *Hypothesis*, as it is more * largely set down by the *Theorist*.

*Lib. 1. chap. 5.

2. But if the Primitive Earth's being of another Form, does depend upon its rising in such a method as this, as indeed it does; then it could not be of another Form from the present Earth, because it could not rise in such a manner, for several Reasons. As

First, Because *it would have taken too long time in doing it*. A longer time by much, than that Divine Account we have of its Origination, does mention, or will allow. For to say nothing how long the Inferior Earth would have been in forming, by the subsiding of the grossest parts of the *Chaos* to the Centre of it; till which were sunk, the other Sedimentals could

could not so well have separated: And to say nothing how considerable a space of time it would have required, for the Aereal Matter to have clarify'd it self, and to have setled in its proper Region: And to say nothing of the Lastingness of that other Purgation, whereby the liquid part of the *Chaos* would have sent forth its Oyliness to invest the Waters, to receive those Dregs that fell out of the Air: To say nothing of these; how much time must have been spent, in producing the exterior Orb of the Earth, which was to be made up of those terrestrial Particles which fell from above, and rested upon the Oyly out-side of the Deep? Should these fine Particles have showred down, as fast as ever we saw small Rain or Snow do; yet how many Days and Weeks must have passed, before they could have swell'd into so huge a Body as the Earth was at first? I say, as the Earth was at first. For according to the *Hypothesis* now before us, the Primitive Earth was bigger than this: So much bigger, as to take all that space into its Ambit, which reacheth up to the tops of the highest Mountains at least. Yea, if the first Earth did not fill a *much* bigger space than that (as it might do; for according to this *Hypothesis*, we know not how far its Circumference might extend) yet a space *somewhat* bigger it must needs occupy, in regard the Mountains are now worn (*a*) *lower* than they were. And for such inconceivable Quantities of little Particles, to descend out of the Air, as would be sufficient to make such a bulky Globe as the Primitive Earth, must necessarily be a *good whiles* work. And so it is expresly acknowledged to be (*Theor.* p. 58, 59.) And then if as fast as they showred down into the

H Oyly

(*a*) Καὶ τῶν ὀρῶν δὲ ὑψηλότατα ἀκούομεν μειούμενα καὶ ἐλάττονα. Τὴν γοῦν Αἴτνην φασὶν οἱ πλέοντες, ἐξ ἐλάττονος ὁρᾶν, ἢ πρὸ τοῦ ἔβλεπον. Τὸ δὲ αὐτὸ καὶ τὸν Παρνασὸν παθεῖν, καὶ Ὄλυμπον τὸν Πιερικόν. *Ælian. Var. Hist. lib.* 8. *cap.* 11.

Oyly Subſtance, they did immediately mix and incorporate with it; yet then it would take up ſome time again, to dry and harden this new made Earth, and to reduce it to an habitable Conſiſtency. And therefore its Formation this way, could by no means fall in with the *real* time of its Production. Nay, it could not be compleated in that ſpace of time, in which GOD declares that he began and finiſht the whole Creation.

3. For that glorious Work is expreſly limited to *Six Days*. And every Day has its reſpective Task particularly ſpecified and appropriated to it. And however more might be created on ſome Days, than is mentioned, as Angels, Hell, *&c.* yet we may be ſure there was no leſs. Not but that GOD could have done all in one Day, if he had ſo pleaſed, or in one Hour, or Minute: As he could alſo have given Being to the World, many millions of Ages before He did. But it was not his Will that it ſhould exiſt ſooner; and his Will it was, that the Creation of it ſhould be protracted to an *Hexaemeron*, or Six Days Work, and therefore he drew it out to that length. But then when *Philo Judæus*, St. *Auſtin*, and others, teach that the World was created in an *Inſtant*; we have no reaſon, *jurare in verba*, to give up our ſelves to a Belief of their Doctrine. Nor is the Saying of the Son of *Syrach* (ſometimes alleged in Proof of the Opinion) to be at all regarded; *(b) He that liveth for ever, creavit omnia ſimul, created all things together*: As if he created them together *in the ſame moment*. Whereas (beſides that the Book is *Apocryphal*) the *Greek* Copy reads it, ἔκτισε τὰ πάντα κοινῇ, *He created all things* in common, as well one as other; in which ſenſe it relates to no time. And accordingly our *Engliſh* is moſt proper,

(b) Ecclus. 18. 1.

proper, *he created all things in general.* Yet this Apocryphal Text seems to be the chief ground upon which St. *Austin* built his Opinion of the World's being made in an instant.

But by that Account which *Moses* gives in, the Earth brought forth Grass, and Herbs, and Trees, on the (c) *Third Day*, even before there was a Sun. Which as it proves that GOD's special Hand was in the Work (and might serve as a Bridle to curb People in their forwardness to idolize the Sun; while it is made plain to them, that he did not, as a DEITY, give Being to these thing by any plastick influence of his, they existing before himself) so it argues withal, that the Earth was then grown into a competent Solidity. Yea, the same *Moses* goes on, and tells the World, That by the end of *three days* more the Earth was made a *Dwelling-place* for Mankind, and in part of it, a *Paradisiacal* one too. And this farther bespeaks it to have arrived at such a Temperament, as it could never have done so soon, in case it had been formed as is above supposed. For besides that the *Theory* allows it to have *become dry by degrees*, after it had *done growing*; it declares, That *the Body or new Concretion* of it was *encreased daily, being fed and supplied both from above and below* (pag. 59.) At which rate, Six Days might easily pass, and Six times Six after them; and the Earth not be fitted for Habitation.

(c) Gen. 1. 12, 13.

4. And though the Sun was not made as yet (the Fourth Day being the first of his existence) yet this does not invalidate *Moses*'s Narrative in the least, by rendring it absurd, or inconsistent with it self, when it tells us that the Earth was brought to such maturity on the *Third Day*. For though there was no Sun then, yet we are assured there was Light: And Pro-

vidence might ſo order this Light, as to have it ſupply the place of the Sun, in meaſuring out Time, and making Days, though not ſo diſtinctly as he does.

And that there ſhould be Light, before there was a perfect Sun, ſeems highly agreeable to the Explication of Light by the *Carteſian* Principles. For according to them, Light conſiſts *in preſſione materiæ cæleſtis*, *in a preſſure of the celeſtial matter*; or, *in conatu ejus ad motum*, in its *tendency* or *nitency to motion*. And therefore if our *Vortex*, or Heaven (made up of this Matter) were by the firſt Mover put into ſuch a Circumgyration at the beginning, as it now has; by virtue of this Gyration, the ſubtile Matter would have been impregnated with a ſtrong *conatus* or propenſity to recede from the Centre of the *Vortex*, according to the Laws of Motion well known in that caſe. And ſo a faint Light would have been caſt through the *Vortex*, at leaſt through the *Ecliptick* parts thereof, though there had been no Sun in the middle of the ſame. And ſo while the Matter for the Body of the Sun had been in preparing (that is, grinding off from the particles of the Matter of the Second Element; which being made to turn upon their own *Axes*, by mutual colliſion, and inceſſant attrition, rounded one another into Spherical Figures) and gathering towards the middle Point of the *Vortex*, and ſetling there; a feeble kind of Light muſt have ſhot from thoſe Central parts of the *Vortex*, till the Sun thus a making could have been finiſht. Which if it might have been in the ſpace of *three* Days (by a great quantity of the pureſt Matter retiring into the heart of the *Vortex*, which ſtill grew bigger, as the Particles of the Second Element had their anguloſities worn off, and ſo grew leſs) then on the *fourth* Day, the Sun might ſhine out in his full ſtrength. While by a

new

new Protrufion, and brisker Propulfion of the Globular Matter, he put a ftronger *conatus* into it than it had before.

And if there could thus have been *Light* antecedently to the Sun; then how there fhould be *Days* and *Nights* at the fame time, is eafie to conceive, admitting the Diurnal Motion of the Earth.

This I have faid, not that I believe the Sun was thus produced (any more than the Great (*a*) Philofopher did) but to make it appear, that the holy Text might be literally true; and that to hold there were Days before there was a Sun, is fo far from being vulgar and ridiculous (as fome would make it) that it is greatly confonant to that which is counted the beft Philofophy. The Noble *Des Cartes*, fo juftly admired, fpeaks the fame fenfe, though in different words; (*b*) *And which peradventure will feem a Paradox to many, all thefe things* (the Properties of Light) *would be juft thus with the celeftial Matter, though there were no manner of force in the Sun, or in any Star about which it is wheeled. So that if the Body of the Sun were nothing elfe but an empty Space, though his Light indeed might not be fo ftrong, yet in other refpects we fhould fee it no otherwife than now we do, at leaft where the Matter of the Heaven circulates.*

As many as are not pleafed with this, have liberty to imagine, that by the three firft (anticipative) Days, more early than the Sun, was only meant fuch certain fpaces of time, as were commenfurate or equivalent to

(*a*) *Quinimo etiam, ad res naturales melius explicandas, earum caufas altius repetam, quàm ipfas unquam extitiffe exiftimem. Non enim dubium eft, quin mundus ab initio fuerit creatus cum omni perfectione fuâ, ita ut in eo & Sol, & Terra, & Luna, & Stellæ extiterint.* Des Cart. Princip. par. 3 fect. 45.

(*b*) *Quodque fortè paradoxum multis videbitur, hæc omnia ita fe haberent in materia cœlefti, etiamfi nulla planè effet vis in Sole, aliove aftro circa quod gyratur; adeo ut, fi corpus Solis nihil aliud effet quàm fpatium vacuum, nihilominus ejus lumen, non quidem tam forte, fed quantum ad reliqua non aliter quam nunc cerneremus, faltem in circulo fecundùm quem materia cœli movetur.* Id. ib. fect. 64.

to three Days; though they were not divided into Diurnal Periods, nor otherwise distinguish't, than by those Acts of Creation which GOD exerted, or the several Creatures which he formed upon each of them. So about the *Pole*, where the Sun is in the *Horison* by Months together, and then out of it as long; Men may reckon the time by *Days*, though they have them not, without any Solecism in their way of compute. But not to dwell upon this Point; they who believe and consider, that there was once an universal darkness, ἐπὶ πᾶσαν τὴν γῆν, (a) *over all the Earth*, for the space of three Hours, a long time after the Sun was made; may, I think, be persuaded to believe also, that by some means or other there might be three real and distinct Days in the World, before he was created.

(a) St. *Mat.* 27. 45. Why it should be read, *over all the LAND*, that is, *Palestine*, as if the darkness had extended no farther, may well be made a question; when it is known that it reached into other Countries. *Dionysius* (to give one Instance) observed it in *Egypt*, being then an Heathen. And is said by *Suidas* (upon his observation of it) thus to express himself; ἢ τὸ Θεῖον πάσχει, ἢ τῷ πάσχοντι συμπάσχει.

5. But that the whole Creation, and consequently the whole Earth, was consummated in *six days*, may be proved by one very good Argument, the *Fourth Precept* in the most sacred Decalogue: *Remember that thou keep holy the Sabbath-day. Six days shalt thou labour, and do all that thou hast to do; but the seventh day is the Sabbath of the LORD thy GOD. In it thou shalt do no manner of work, thou, and thy son, and thy daughter, thy man-servant, and thy maid-servant, thy cattel, and the stranger that is within thy gates. For in six days the LORD made heaven and earth, the sea, and all that in them is, and rested the seventh day. Wherefore the LORD blessed the seventh day, and hallowed it.* Now the *Seventh* day, which by divine

Bene-

Benediction, and special Consecration, was set apart for the *Jewish Sabbath*, was no other than a *Natural* Day, consisting of Twenty four Hours. And the *six Days* in which God allowed Men to work, were of the same quality or duration. But then he permitting labour six Days in the Week, because in six Days he made Heaven and Earth, the Sea, and all that in them is; and enjoyning a cessation from all manner of work, and a Sanctification of the Seventh Day, because on it he rested from his work of Creation; from hence it will follow by undeniable consequence, that the *seventh Day* on which GOD rested, and the *six Days* on which he wrought, must be of the same nature that the *Jewish* Sabbath and Week-days were of; and that in *Six* such Νυχθήμερα, or *Natural* Days, consisting of Twenty four Hours apiece, the Heaven, the Earth, the Sea, and all that in them is, were made.

And indeed unless the HOLY GHOST had said expresly that they were *Natural* Days, each of them made up of Twenty four Hours, he could hardly have spoken them to be *such*, more plainly and properly than he has done. For he tells us *seven* times over (in the First of *Genesis*) that is, concerning every one of the seven Days, That they had *Evening* and *Morning* (ויהי ערב ויהי בקר יום.) Whereas if he had made use of the single word יום, *that* might have been taken for *time* indefinitely or at large. But as if he designed to prevent this, or at least to give no occasion for it; as often as he mentioned יום he was still pleased to tack ערב and בקר to it; to evidence that he meant no other than a *Natural* Day, to which *Evening* and *Morning* do belong.

And that which makes it farther evident, that he has tied up יום *Day*, to signifie a *Natural* Day, in the

the Story of the Creation, is, That this *Fourth Commandment* is partly entred in that Story; and יום השביעי *the Seventh Day*, which he blessed and sanctified for the *Sabbath*, is said to be the seventh Day on which he rested from his works. And so both were Natural Days alike, the one as much as the other. And therefore the Seventh Day separated for a Sabbath, and kept so by the *Jews* (from the Promulgation of the Law, to the Dissolution of their Polity, yea, to this very Hour, in their Dispersion) being a true *Natural* Day, the *Seventh* Day whereon GOD rested from his works must be the same.

And truly once to pretend that there is any thing of Cloudiness or Ambiguity in the recited Precept; or that GOD by the Days mentioned therein, did not mean ordinary *Natural* Days; would be to raise a mist to darken the Truth; to offer to tye a knot, where there really is none; and to put plain words *by* their common sense, meerly to force a difficulty into them. Suppose a Man should command his Servants or Children to work six Days, because he himself in six Days had done such and such things; and to rest on the seventh Day, because on that Day he ceased from his labours: Could it ever enter into the thoughts of any, but that the Days in which he wrought, and they are to work; and the Day in which he rested, and they are to rest, must be of the same nature? Why, such is the case here, if we put but the Great GOD into the place of that Man, and allow his Precept to be of a plain import and signification, as we have reason to do. For this Great God was now publishing a Law to his People: A Law whereby they were to live, or die for ever: A Law by which he really intended, and heartily desired, that they should not die, but live to Eternity. And

He

He being in hand with a Law of ſuch conſequence to their pretious Souls, who can queſtion but it was worded plainly? For without doubt he would deliver it in ſuch familiar terms, as might be moſt intelligible to the meaneſt Capacities amongſt that People, to whom he recommended it as a ſtanding Rule. So that the Fourth Commandment being a piece of that Rule, a Branch of that Law, we cannot ſuppoſe it to be cloathed with obſcurity, either in the ſubſtance or reaſon of it. And truly if ſo plain a Paragraph as that be not to be taken in its Grammatical ſenſe, 'tis impoſſible we ſhould know the Mind of Scripture, and vain it will be to pretend to underſtand any Period in it. But then, if where God ſpeaks plainly in his Word, we muſt underſtand it Literally; ſurely we muſt do it moſt of all, in that part of his Word which is the Body of his Law, and ſo the ſpecifick Rule of our Practice. And if the Moral Law in general was of a Literal Signification; then ſo was the *Fourth Word* of it to the *Jews*. And if *that* were Literally to be interpreted, as undoubtedly it was, the World muſt be created in the time there ſpecified, in juſt *Six Days*, that is, neither more nor fewer.

And *Moſes* methinks ſeems to be mightily concern'd to ground Men in this. For though he had noted already in the Firſt of *Geneſis*, that the work of Creation was compleated in Six Days; and had fairly accounted for each of them in particular: yet reviewing things in the Second Chapter, he there inculcates the ſame afreſh, that ſo they might take farther notice of it. *Thus the Heavens and the Earth were finiſhed, and all the Hoſt of them.* That is, in the Days, and according to the order before remembred. But had the Earth been formed after the tenor of this new Hypothesis, it could never have been finisht in *Six Days*, and brought into a condition fit to entertain its Host Creatures, in that compass of time. Unless at the rate of the *Mahometan* Miracle, we should yield that the Showers of Earthly Particles, were hardened by instantaneous induration. For so it is storied of a certain *Dervich*, or monkish *Musselman*, that strowing Sand upon the Waves, as fast as he sprinkled it, it turn'd to a Causey before him, whereon he might walk to a *Mosque*, the usual place of his Devotions. Though far more agreeable to this *Hypothesis* (as it makes the Formation of the Earth so slow) is the Phantasie of the *Chineses*, inhabiting *Formosa* and other Islands. Who hold that the World, when first created, was without form or shape : but by * *Pankun*, one of their Demi-gods, (the *Sixty Second* of their *Seventy Two* Deities) was brought to its full perfection in Four Years.

* *Atlas Chin.* part 2. pag. 46.

6. The only considerable way of eluding this Doctrine of the Worlds being made in *Six Days*; is the introducing of *Cabbalism* into the Story of the Creation. For by *That*, Numbers, which show the order of time are made Types or Emblems, or secret Notes of the Natures of things. But the force of the Holy Argument alleged , is not to be shifted off by this means. The *Jews*, without question, had a *Cabbala* amongst them; and upon this *Cabbala* they set an exceeding high value. For they put it in the Scales (as some Christians now a-days do Tradition) even against the written Word it self. So that known saying witnesseth, דברי קבלה כדברי תורה דמו *the words of the* Cabbala, *and the words of the* Law, *are alike*. Which how well soever it might

might agree with that true and ancient *Cabbala*, which they had from *Moses*, and in part from *Adam*, as † some think; it suits not their present † Reuchlin. de Art. Cabbal. p. 9, &c. *Cabbala* at all. Upon *that*, * *Cunaeus* bestows a mean and disgraceful, though (I suppose) a most * De Rep. Heb. fitting character; where he declares it to be *full of trifles*, &c. Though *Mirandula* on the other side, crys it up as much (if the same) in his *Apology*: where he avers it to have been first committed to writing by *Esdras*. But that which I would note is *this*, That the genuine *Cabbala* of the *Jews*, cannot overthrow that strength of Argument, in the *Fourth Commandment*, for the Worlds being made in the space of *Six Days*. The Reason follows.

7. The Cabbalizing *Jews*, even after the purity of Oral Tradition was lost amongst them, were still hugely careful to deliver nothing, but what was consistent with the Literal sense of Scripture. So that in treating of any Period of it, they did but orderly proceed from one sense to another: and their *Traditional* sense was to be so far from being repugnant or destructive to the *Verbal* one; that it was to comply with it, and to be subordinate to it.

Thus, though (as *Aben-Ezra* says) the Law, besides דרך הפשט *the Literal way*; hath שבעה *seven*, or שבעים פנים *seventy faces*, or ways of Interpretation; and be capable (in places) of *Tropological*, *Allegorical*, *Anagogical*, and *Cabbalistical* meanings and applications: Yet it is a known *Rabbinical* Rule, אין שפת המקרא נופלת על המדרש *The Style of the Scripture falls not in with the Midrash*, or Allegorical Exposition. Which is seconded with another of the like import, אין מקרא יוצא מידי פשוטו ומשמעו *The Scripture departs*

departs not from its simple and literal meaning. It was delivered by that eminent Man for Learning, *R. Solomon Jarchi (Præfat in Cantic.)* Not that he meant Sripture is *always* to be kept to a *literal* ſenſe only; the very words wherewith he uſhers in that Rule, ſufficiently declare the contrary; מקרא אחד יוצא לכמה טעמים *One Scripture may be drawn out into many ſenſes.* And therefore he only meant, that the *literal ſenſe* of Scripture is not to be thrown off, or taken away, by, or for *any other.* And *R. Moſes bar Nachman* lays down the ſame Rule. To which ſeveral other *Rabbies* alſo do conſent; particularly *Menaſſeh ben Iſrael,* who in his Firſt Book concerning the *Reſurrection of the Dead,* ſays, *That Scripture is always to be underſtood and explained according to the* Letter, *unleſs where ſuch an Explication implies a contradiction.* According to which *Axioms* of the Learned, what ever ſenſe might be ſuperinduc'd to the Text, they were ſtill to adhere (ſo far as Reaſon would ſuffer) to the *literal* one. Yet where they were reconcilable, and could fairly be coincident, they were both allowed. And therefore when in Controverſie, ſome have been high for the *Letter*; and others hot for the Allegory or *Cabbala*; the difference betwixt them has been frequently accommodated, by pronouncing that *Talmudic* Sentence, יש אם למקרא ויש אם למסורת. By which conceſſion of a complicated ſenſe in the word or clauſe, that miniſtred occaſion of fierce Diſpute, the contending Parties have both been quickly and fully ſatisfied. And well they might; for by the eaſie Umpirage of that ſoft and moderative Saying, they were immediately brought to equal terms, and ſet, as I may ſay, upon even ground. For each of them had warranty to maintain their own Opinion, and liberty to acquieſce therein reſpectively. Though how much, מקרא, the *common Lection* or Senſe, was preferred

preferred before, מסורת, the *Traditional* one (when they came in competition) is clear from a Passage in the learned *Buxtorf*'s Treatise *de Punct. Antiq.* pag. 104. where speaking of the *Talmudic Axiom* now cited, he says, *Quando hoc priori contrariatur, & illud evertit, nulla ejus habetur ratio, sed fundamentum manet in* מקרא. *When the* Cabbalistic Exposition *is contrary to the literal or received sense, no account is made of it, but the* foundation *rests in the common sense or reading. Quando verò ei subordinatur, &c. But when it is subordinate to it, and consistent with it, and* both *may be reconciled; they admit* both.

And such a mixt or compound sense of Letter and *Cabbala*, well consistent with one another, they believed to be in the *Mosaic* Cosmology, or the Story of the World's Creation. Besides its common, humble, obvious sense, which lies bare to view, and offers it self freely at first sight; they moreover conceived it big with another more occult and remote, and also more sublime than that; where the Kernel of Mystery lay close shut up, *sub cortice verborum*, under the Husk of plain and ordinary words. And for this reason they used to restrain Men from reading it, till they came to maturity both of Years and Judgment, as St. *Jerom* testifies, in the Preface to his Commentaries on *Ezekiel*; * *Unless one has attained to the Age of Priests when they enter on their Ministry, that is, to thirty years, he is not suffered, with them, to read the Beginning of* Genesis, *&c.* But then if the *Jews*, together with their *Mysterious*, always held a *literal* sense in Scripture (where it can take place) and also kept this Mysterious sense of theirs from breaking in upon the Literal, and doing violence to it; according to their own recited Rules: from hence it is manifest, that the World might be Created in just *six Days*,

* *Nisi quis apud eos ætatem Sacerdotalis ministerii, i. e. tricesimum annum impleverit, nec principium* Geneseos — *legere permittitur.*

for all their *Cabbala*, or the numbers in it. And when *they* kept the *Cabbala* within such bounds, *Christians* have no reason to stretch it farther; for they have it but from *them*, or rather (as to the *Numeral* part of it) from one that had it (or is presumed to have had it) from them, I mean, *Pythagoras*.

8. But grant (what by no means is to be granted) that there were a *Cabbalistic* sense in the Story of the Creation, so venerable and excellent, as to be allowed to supersede the *literal* one quite, or to swallow up the same; and that the *Mosaic Numbers* there, were not at all intended to distinguish *time*, but only to shadow out the *Natures* of things; and so no satisfactory Proof could be fetcht from thence, of the World's being compleated in *six Days*: Yet still the Holy Argument produced, would hold good; because in that Divine Precept there can be no *Cabbalism* exclusive of its *literal* meaning. And that for these Three Reasons.

First, *Because GOD must then have put dark Mystery into the heart of his Law.* Into that part of it which (that it might be sure to be most plain, as well as authentick) he was pleased to write with his own Finger. Into that part of it which every poor *Israelite* newly come out of Egyptian Bondage, was to practise. And therefore it was necessary he should understand it, and consequently that there should not be obscurity in it: A thing ever held incongruous to Laws (as very unsuitable to their Use and End) and always declined by wise Legislators. For they have still been careful that the Statutes drawn up and enacted by them, should carry a clear sense along with them both in their Injunctions, and the Reasons of them.

Secondly,

Secondly, *Because then something of Duty expressed in the Commandment, would have been very improperly and incompetently urged.* That Men should work six Days in the Week, because in six Days GOD created the World, upon each of them bringing some considerable Pieces of it into being, is most fit and reasonable: Even as fit and reasonable as it is for mean Persons to imitate their Superior; for Creatures to follow the great Example of their Maker. But that Men should labour six Days in the Week, because the Number *Six* is the Character of the Nature of any Creatures, would be altogether empty, trifling and impertinent, as admitting no manner of dependance or connection betwixt the Reason and the Thing.

Thirdly, *Because it would shuffle and jumble the Natures of things together, or else bring a strange Confusion into the Numbers of the Cabbala.* It would *shuffle the Natures of things together*, in an intolerable manner. For thus the Heavens and the Earth (as the Commandment runs) the Sea, and all that in them is (which according to the Physical *Cabbala* of *Moses*, are thought to have their Orders, and various Natures, distinctly characterized by several Numbers) must be here referred to *one single* Number *Six*. And *That*, according to other Numbers (for why should it be of a more dilated Mystick significancy than the rest?) mysteriously pointing out but *one single* Nature; to refer all things to that *one Number* (as we must do, if we put *Cabbalism* into the Fourth Commandment) would be to reduce them to *one single Nature*: I mean, to the *Animal* one; to signifying of which the Number, *Six*, is both naturally inclin'd, and also determin'd by its Denominations.

First,

First, It is Naturally inclined that way. For it is said by *Cabbalists*, to be made by drawing (*a*) the *Masculine*, *Three*, into the first *Feminine* Number, *Two*. For *three* times *two* is *six*: And then *six* into *six*, is *thirty six*. One way it points out *Male* and *Female*, and the other way *Procreation*, and both ways it clearly relates to *Animals*.

(*a*) Ἄῤῥην τε καὶ θῆλυς ἐκ πέφυκε, κατὰ τὴν ἑκατέρου δυνάμεως ἡρμόσθη, &c. *Philo in lib. de Mund. Opif.*

Secondly, It is determin'd the same way, by its *Titles* or Denominations. They tye it close and fast to such meanings, as all-along restrain it to the Properties or Operations of the Animal Nature or Condition. Thus it has the Names of Φιλίωσις and Φιλοτησία· because Animals of the same *Species* have a sort of love or friendship for one another. Of Ζυγία and Ζυγῖτις· because they yoke or joyn themselves together in Pairs, *&c.* It is also called Ἀφροδίτη and Ἄκμων· because of their venereal Copulations; and their continual Productions, by a never-failing constancy of Propagation. Διάρθρωσις· because their Bodies are full of Articulations or Joynts. Κόσμος, Ἀλήθεια and Ἁρμονία· because of that comely Structure, true Symmetry, and admirably curious and useful Contexture of their several Parts and Members. Πανάρκεια and Ὑγεία because of that sound constitution which Providence endu'd them with, and power sufficient to keep up their respective Kinds in the World. The like may be said of Γάμος, Γαμήλια, Ἀνδρογυναία, Ψυχογονία, &c.

Vid. Nicomach. Gerasen. Arithmet. Theolog. lib. 2.

Meurs. Denar. Pythag.

And when the *Senary*, by its Nature and Appellatives, is thus particularly appropriate to the Denotation of the Animal part of the World; if in the Fourth Commandment we take it in a *Cabbalistic* sense, we must suppose the things there referred to it; that is to say, the Heaven, and Earth, and Sea, and all that in them is, to be living Creatures; which

which would intollerably confound the Natures of Beings.

Or elſe, which would *bring ſtrange confuſion into the Numbers of the* Cabala ; this one, the *Senary*, muſt in power be equal to all the reſt. That is, it muſt denote what the *Unite*, *Binary*, *Ternary*, &c. are ſuppoſed to do ; and ſo ſignifie Spiritual, and Material Beings ; Terreſtrial, and Aquatick Animals ; Animate, and Inanimate things, all at once. But to ſtretch it out to ſuch a comprehenſive Symbolicalneſs, as to make it an Hieroglyphick of the Univerſe, and of all things in it: might ſeem to be a violent ſtraining it, beyond its juſt Latitude. Eſpecially if we conſider, that the reſt of the Numbers before it, are all conceived to be fitted and reſtrained, to the ſhadowing out ſuch and ſuch *ſingle* Natures: and none of them allowed to be of ſo wide a ſignificancy, as to have a Myſtical reference to all the Creatures.

Though truly if any witty buſie *Cabbaliſts*, ſhould either out of the Nature or Names of *this* Number ; extort ſuch meanings, as might make it ſeem an Emblem of the whole Creation, as to its production, compleatment, or the like : we need not be much ſurprized at it. For no wonder that they ſhould ſo vex (as I may ſay) and Wire-draw Numbers, as to force and wind them even to what they pleaſe ; when they have power to put any commune word upon the Rack, and to torment it at ſuch a rate, as to make it ſay what they liſt. *Mirandula* proves this by a pregnant Inſtance. For in his Expoſition of the word בראשית (at the end of his *Heptaplus*) which ſignifies no more than , *in the beginning* : he does ſo toſs and tranſpoſe the ſeveral Letters and Syllables of it ; and ſo ranſack, and ſqueeze, and

K torture

torture it; as to make it ſpeak all this long Sentence: *Pater in Filio & per Filium principium & finem ſive quietem creavit caput ignem & fundamentum magni hominis fœdere bono.* They that would know the ſenſe of it, let them conſult the Author. By it he deſigns, *guſtum dare* Moſaicæ *profunditatis, To give a taſte of* Moſes'*s profoundneſs.* But ſhould any attempt (in imitation of him) to interpret *Moſes*'s whole Story of the Creation; as he has done the firſt word of it: they would certainly find *Moſes* to be profound indeed; and themſelves to have lanched out into ſuch a deep, as would prove to have neither bottom nor bounds in it. And yet the Noble *Picus*'s way, is but *one* of the *three* ways of Cabbalizing by *Permutation*: and *Permutation* it ſelf, or the *Tranſpoſing* of words, is but *one Sort* of the *Speculative Cabbala* neither. For beſides תמורה there are other two kinds of it, גמטריא, and נוטריקון. So that according as *Cabbaliſts* have been, either for laudable phantaſie and ingenuity; or elſe for humor and extravagance: they have had as large Fields to expatiate in, as they could deſire. But I muſt not digreſs too far.

By what has been ſaid, I hope it is evident, that in the Fourth Commandment, there can be no *Cabbaliſtick* meaning, at leaſt none *deſtructive* to the *literal* one. Whence it will follow, that the whole Creation was begun, carried on, and conſummated, in *Six Days*: the thing I was to prove. So that if the *Theoriſt*'s conceit touching the Earths Formation be true; *Moſes*'s Account of the Creation muſt be falſe. And the Hiſtory he wrote of it, as the unerring Pen-man of the HOLY GHOST; inſtead of being of unſuſpected Credit and Authority, we may juſtly call (as *Celſus* did of old) * *μῦθόν τινα ὡς γραῶν, an old Womans Fable.*

* Orig. *cont. Celſ. lib.* 4.

9. It

9. It is manifest also that *Moses* in that sacred Story, did not make use of *meer Numbers*, but of *Time*: yea, of time divided into known and common Periods; namely, into *Days*. And those *Days* (which is especially to be noted) are expressed (as was said before, and must always be remembred) by *Evenings* and *Mornings*; just as the *Hebrews* speak *Natural* Days. And therefore not to allow the *First*, *Second*, and *Third* Day, *&c.* there mentioned; to be such *vespero-matutina*, *Natural* Days having Evenings and Mornings: But to turn them into pure *Numbers*, *One*, *Two*, *Three*, &c. is highly unwarrantable; as being a double injury to Scripture, upon no necessary account. For it is a depriving it of its direct and genuine sense; and a forcing it upon another quite beyond and beside that, only to gratifie the humour of *Cabbalists*. Which humour or phantasie of theirs, many times is vain and trivial, and grounded upon nothing perhaps, but ignorance or mistake.

To evince as much, let me give in one Instance of this nature, most pertinent to the matter we have in hand. *Philo* (that mighty Man among the *Jews* for allegorizing the Story of the Creation, and to whom later *Cabbalists* are greatly beholden) positively and aloud pronounceth thus, (*a*) *It is a silly thing, to think that the World was made in Six Days, or in any certain time.* And why so? His reason in short is, *Because it was made before there was a Sun*, and so there could be neither *Time*, nor *Days*. But there might be Days, before there was a Sun (as we have shewed) and so where's the strength of that Objection? In like manner when it is said, *In the*

(*a*) Εὔηθες πάνυ τὸ οἴεσθαι ἓξ ἡμέραις, ἢ καθόλου χρόνῳ κόσμον γεγονέναι. *Leg. Allegor.* lib. 1.

BEGINNING GOD created the Heavens and the Earth: he will not allow that, BEGINNING, should signifie *time*; for a like reason. *(a) Because time was not before the World, but was made either with, or after it.* Strangely argued, for so Learned a Man. As if the World could not have been made *in the beginning* of time, because time was made *with* it. Whereas if it had not been made with it, if it had not began just then; the World could not have been made *in the beginning of it* indeed. Nor does he back this reason at a better rate, where he adds; *For (b) seeing that time is a space of motion of the Heavens, motion cannot be before the thing moved.* Yet let us but suppose, that the World was made, and the Heavens put into motion *at once*; and then the World would be created, when time began: as being created together with that motion in which consists the nature or measure of it. And yet he concludes for himself, as if he had argued most cogently; *But (c) if so be,* BEGINNING, *be not taken now according to time; 'tis fit it should be taken according to* Number. And so (in part) we have an account, how *Numeral Cabbalism* crept into the Divine Text of *Moses.* Even because Men had not Philosophy enough to make out, how time and days might be before the Sun. But is not natural Philosophy then an *useful* thing, and of *great use,* according to its Character, in the First Chapter?

(a) Χρόνος γὰρ οὐκ ἦν πρὸ κόσμου, ἀλλ' ἢ σὺν αὐτῷ γέγονεν, ἢ μετ' αὐτόν. *de Mund. Opif.*

(b) Ἐπειδὴ γὰρ διάστημα τῆς τοῦ οὐρανοῦ κινήσεώς ἐστιν ὁ χρόνος, προτέρα τοῦ κινουμένου κίνησις οὐκ ἂν γένοιτο. *Ib.*

(c) Εἰ δ' ἀρχὴ μὴ παραλαμβάνεται τανῦν ἡ κατὰ χρόνον, εἰκὸς ἂν εἴη μηνύεσθαι τὴν κατ' ἀριθμόν. *Ibid.*

10. Were

10. Were Enquiry made how the Church of CHRIST hath resented *Calbalism*; or what respect she hath shown to such as used it: an Answer might partly be shaped out thereunto, from her Carriage towards *Origen*. That Father had a peculiar Talent (above others) at Allegorizing Scripture, and in delivering Doctrines of the *Cabbalistic* strain. But how did Holy Church receive his Notions of that stamp, and how did she deal with him for their sakes? This we may learn from * *Photius*, who tells us, That the Fifth Universal Synod, κατεδίκασι καὶ ἀνεθεμάτισεν Ὠριγένην, *condemned* Origen, *and anathematiz'd him*. And for what cause? Why, for that he attempted to introduce Ἑλληνικὴν μυθολογίαν, *the Greek Mythology*, or Pagan Fables, into the Church of GOD. And particularly for that piece of *Dotage* (it is the Patriarch's word) in teaching προϋπάρχειν τὰς ψυχὰς τῶν σωμάτων, that humane *Souls were preexistent to their Bodies*. And as all know, *preexistence* was a principal Branch of the *Cabbala*.

* *Epist. ad Mich. Bulg. Princip.*

11. And truly the *Cabbala*, which makes Numbers emblematize the Natures of things, may well be rejected. For indeed it proclaims its own *vanity*, in one notable Instance: I mean, in the *Coincidence* of its Numbers, as to their *symbolical significancy*. In the whole Story of the Creation, there are but *Seven* Numbers made use of. Now if GOD, or *Moses*, had designed these Numbers for a Mysterious use; we need not question but care would have been taken, that *Two* of them should not be Symbols of the same thing, when *One* would have served every whit as well. For so one of the Two Numbers would be superfluous: Yet supposing

posing that there is such a *Cabbala* as some contend for, in the Story of the Creation; there must be this vanity or superfluity in it. For then the Number *Five*, and the Number *Six*, in their Mystical Property, must refer to one and the same thing, *viz.* the *Animal Nature*.

How the *Senary* is an Emblem of that, both by its *Nature* and its *Names*, we have seen already. And truly the *Quinary* is made a Symbol of the same, and that *both* those ways. First by its *Nature*. It consists of the *Masculine* Number, *Three*; and of the *Feminine*, Number, *Two*: and so it mystically signifies *Male* and *Female*. *Five*, also drawn into *Five*, brings about *Five* again; *Five times five*, is *Twenty five*: so it betokens *Generation*. And *Male* and *Female*, and *Generation* by them, we know, relate directly to *Animals*. And then for *Names*, it has *Cytherea*, and Γάμος, which are Names of the *Senary*; and so it must be of the same mystick signification with that still.

Nor can it be pretended (to diversifie the mystick significancy of the Numbers) that the *Quinary* refers to *Water Animals* only. For not only *Fishes* were made on the *Fifth* Day; but *Fowls* too: and that *out of the ground* every one of them, *Gen.* 2. 19. And as no Fowls live altogether *in* the Water; so very *few* kinds of them live *upon* it, in comparison of those that do not. In all respects therefore the *Quinary* seems to be a meer supernumerary. It signifies nothing, but what the *Senary* could have signified as well.

Whence we may conclude, That either *Moses* was guilty of a notorious bungling Oversight, in inserting an useless number into the *Cabbala*; which in so rare a Philosopher as he was, must not be admitted;

mitted; eſpecially he being divinely inſpired, and ſtudiouſly contriving ſo profound and admirable a piece of Myſtery, as the *Cabbala* is reputed: or elſe (which is the truth) that this very thing does betray the vanity of the *Cabbala*, and ſhows it to be but an ingenious Phantaſie.

12. Now the *Cabbala* being a thing ſo improbable; and the *Literal Senſe* of Scripture ſo very authentick, as not to be thrown off, or put by for any other, where it can be held to: it remains that the Story of the Creation is to be underſtood according to that ſenſe. And ſo where *The Theory of the Earth* is contrary to that ſenſe, or not agreeing with it; it is to be look't upon all the way, as contrary to Scripture, and diſagreeing with *That*. Not that I deny there is Myſtery in the Story of the Creation; for undoubtedly there is a great deal, and that ſo deep, that it is hard to ſee to the bottom of it. But once again I ſay, (to prevent my being miſtaken) That no *Myſtical* or *Cabbaliſtic* ſenſe is to be approved of, that overthrows or nulls the *Literal* one. And the reaſon is plain; becauſe if the *Literal* ſenſe ſhould be taken away, it would ceaſe to be an Hiſtory: and alſo could have nothing of fixed or certain meaning in it; but might be moulded any way, and changed into every thing, according to the various apprehenſions of Men and their working Phantaſies.

13. Let me here caſt in this as an Overplus. Where Scripture delivers it ſelf ſo as not to be literally interpreted; it is *ſometimes* done out of greater plainneſs that it affects, and the better to accommodate it ſelf to our capacity. Thus when it expreſ-

ſeth

ſeth GOD's power, or his doing any thing; by *Hands*. His Knowledge, or Obſervation of any thing; by *Eyes*, &c. It is meerly in way of Condeſcention to us, and to render what it ſpeaks of, more eaſie and familiar to our apprehenſion. Here therefore that Axiome of the *Talmudiſts*, remembred by great *Maimonides*, takes place, דברה תורה כלשו בני אדם * *The Law ſpeaks according to the Language of the Children of Men*. And for this reaſon (I think) the Book of *Canticles* is ſo parabolical and alluſive. Not to veil and darken the ſenſe, but the better to illuſtrate its Divine Argument: and the more fully and fairly to ſet forth that paſſionate affection and dearneſs, which is betwixt the moſt GLORIOUS JESUS, that great Lover of Souls; and all zealouſly religious and devout Perſons.

* Mor. Nevoch.

CHAP.

CHAP. III.

1. A *Second Exception* against the *Formation of the Earth*, viz. the *Fluctuation of the Waters of the Chaos, whereon it was to be raised.* 2. That *Fluctuation* caused by the *Moon.* 3. The *Theorist's Doubt*, whether *she* were then in our Neighbourhood, considered. 4. The *Precariousness* of his *Hypothesis* in several things relating to the *Chaos*: *Which* ought to have been *better cleared and confirmed*, according to his *own declared Judgment.* 5. The Descent of the Earthly Particles out of the Air, not only Precarious, but *Unphilosophical.* 6. And also *Anti-Scriptural.*

1. AS every Building must have a sutable Foundation, so fit it is that the Earth should have such a one; It being not only a stately Fabrick in it self, but moreover designed, at the formation of it, to be the Mansion or Dwelling of a World of Creatures. And it being destined to so great and noble an use, what pity had it been that it should have miscarried in the making, and have sunk into Ruines, while it was setting up, for want of a sufficient foundation to support it? Yet had it been built after the manner aforesaid, perhaps there would have been no less defect in the Architecture of it, than the want of a meet Foundation. For it was to be reared upon the Waters risen out of the *Chaos*; and were they fit to bear such a mighty Pile? I mean in regard of their unsteadiness and constant Fluctuation.

That the Earth might be ſpread, and by degrees raiſed upon them; Was it not requiſite that they ſhould be of a quiet and even ſurface? Otherwiſe it may be the unctious Subſtance could not have gathered upon *them*, nor the Earthly Particles have ſettled upon *that*. But the incontinuity of the one (it being broken by the motion of the Waters) leaving many open ſpaces to the other; through thoſe ſpaces they would have ſunk right down to the bottom of the Deep, and no Earth would have been produced.

2. Yet that thoſe Waters were quieſcent and even, upon due examination will hardly be found. For the *Chaos* in the beginning was turned about upon its own Center; elſe how comes the Earth to be ſo now? And if it was carried about by ſuch a Gyration, how could the face of its Waters be ſtill and equable? Not that I mean they were diſturbed by the meer Rotation of the *Chaos* neither; but by ſomething elſe in conjunction with *that*, namely, The bulky preſence of the Moon. For if ſhe was then placed, where now ſhe is, what hindred but ſhe might have the ſame Motion which now ſhe has? And if ſhe moved then in an *Eliptic* Circle about the *Chaos*, as now ſhe does about the Earth (as in all reaſon ſhe ſhould; the *Chaos* being then ſituate, where the Earth is now, betwixt the Heavens of *Venus* and *Mars*; from which ſituation that *Eliptic* Circle reſults) would not the Waters have been too much diſcompoſed thereby, to have been a fit Foundation for the primitive Earth.

Indeed they being in all places of an equal depth, and flowing freely without reſiſtance; it is very probable that the Tides then were leſs fierce and rough, than

than they are now. And yet they all making but one Sea, and that being open and exposed to the Moon; it is as probable again, that they swelled extreamly, and went mighty high. Now the Moon squeezing them her self (by streightning the Heavens) on one side of the *Chaos*, wherever she was (as at her *Zenith* suppose) and occasioning a like compression of them on the opposite side (or at her *Nadir*:) and the *Chaos* still turning round upon its own *axis* every Four and Twenty hours: from hence it follows that these Waters felt the force of two Tides, in every such space of time. Now where they Ebbed end Flowed so frequently and incessantly; must not their Æstuation have been so turbulent, as to have hindred the gathering, or dissolved the continuity of the Oily Matter; and so have prevented the Earth's superstruction upon it?

In case it be urged, That the Unctious Matter upon the face of the Waters, was so very thick, as that they might heave and sink under it, as there was occasion, and never break it: I answer, When this Oily Substance did first arise, it must needs be thin, and so apt to be broken and divided; and *that* being disjoined, the Earthly Particles falling in at the void spaces, would have sunk directly (as was said even now) through the Waters, having nothing to support them. And then (which is farther considerable) the heaviest Particles of Earth, descending at the same time, in far greatest plenty (the Air being then fullest of impurity, and purging it self most freely) they would have come down so fast, and in so great abundance, as easily to have overpowered the thin Oily Scum on which they fell; and being a little soaked in it, and incorporate with it, have weighed it down in Flakes to the bottom of the Waters; upon the top of which it could no longer float, as being overloaded with the heavineſs of the imbodied Earth. And truly the flowing of the Waters with a ſtrong head now this way; and their returning by and by with as much force the contrary way: muſt needs put them into ſuch reſtleſs agitations and croſs commotions, as would have much promoted the diving of the Flakes aforeſaid.

Nor are we to meaſure the motion of the Chaotic Waters, from the preſent great Seas. For however *they* may be leſs diſcompoſed by Tides, yet nature *then*, was in other circumſtances (according to the *Theory*) than it is *now*; and thoſe Waters might be moved at another rate, than theſe are. For our preſent Earth was at that time all diſperſed in the Air. And the thicker and fuller the Air was, the ſtronger preſſure would the Moon make upon that; and that again upon the *Superficies* of the Waters: and conſequently the higher muſt the Tides riſe, and the more violent muſt they be.

And then the *Theory* makes another motion in the Chaotick Waters neceſſary, namely, A Defluxion of them from the *Æquator* towards the *Sides* or *Poles* of the liquid Globe; in order to the forming it (and conſequently the Earth to be raiſed on it) into an *Oval Figure*. And this motion might create a new diſturbance in that Element. Yea, not only ſo, but it might moreover be fatal to the riſe of the Earth. For (*a*) *the watry Globe was to grow oblong, by the flowing down of the Waters to the ſides* (they are the words of the *Theory*) *and the disburthening the middle parts about the* Æquator. But then when

(*a*) *Ex illo defluxu aquarum ad latera, & exoneratione partium mediarum circa Æquatorem, Globus aqueus deveniret aliquantulum oblongus.* Theor. pag. 198.

theſe

these Waters did thus recede or discharge themselves from about the *Æquator* or middle of the Globe, and flow down to the sides of it; how easily might the Oily Matter have followed their course? Yea, perhaps how necessary was it for it to do so? While the uppermost Waters thereabouts being most hurried, and most at liberty; would have fallen back, and carried that away with them. But then if the upper Waters thus drew off, and the Oily substance slid away upon them, what foundation could the Earth have had in those middle parts we speak of? Especially if these Waters continued their course for any time; as it was needful they should to bring about the effect mentioned. For so vast a body of Waters, as that of the Abyss, could not by this means, of a perfectly *round*, be made into an *oval* or *oblong* Figure, on a sudden.

3. But in reference to this matter, there is a *Doubt* made by the *Theorist*, which must be considered and removed. Otherwise most of what has been said, touching the instability and fluctuation of these Waters, will be vain and groundless. The *Doubt* is, (a) *Whether the Moon were then in our neighbourhood.* And truly I had almost said, he might next have questioned, *whether the Sun were then in our Heaven*: there being in the Story of the Creation, no better evidence for the one, than for the other. I confess the suggestion (as wild as it is) would have done the *Arcadians* a great kindness. For they used to boast of (what was always a Riddle and nonsense to the Wise) their being more ancient than *Jupiter* and the *Moon*. So says *Ovid*:

(a) Theor. pag. 241.

Ante

Ante Jovem *Genitum Terras habuisse feruntur*
Arcades ; *& Lunâ Gens prior illa fuit.*

But the service it might have done them, as to this arrogant brag, will by no means countervail that dammage which it does to the person who raises the Doubt. For it involves him in the guilt of unhappy temerity towards the Holy Writings: Yet the *Theorist* does not only start this Scruple, but argues for it thus, (*b*) *Her presence seems to have been less needful; when there were no long Winter-nights, nor the great Pool of the Sea to move or govern.* Too bold an affront to Scripture.

(*b*) Ibid.

That says expresly, That GOD made TWO great Lights; and both upon the Fourth Day, *Gen.* 1. 16. The *Theorist* suspects he made but *One.* And truly let him but allow *Two* to be made, and the *Moon* of necessity must be come into our Neighbourhood ; because she alone could be a *Great* light in the neighbouring Heaven, to make up the Sun, *Two.* There is no bringing any *Star* into the Number. For though the smallest of them be a *truer* and *greater* Light than the *Moon*; yet no one of them, was ever a *great* Light in this lower World: and GOD created more than *Two* such. Besides, Scripture says, That when GOD made two great Lights, he *set them* (*both* of them, both of them *then* on the *same day*) *in the Firmament of the Heaven, to give light upon the Earth.* And must not both of them then be in our neighbourhood at that time? And lastly, It says, That as GOD made the greater of these Lights to rule the Day; so he made *the lesser to rule the Night.* And when did the lesser begin to rule the Night?

Night? Why, juſt when the greater began to rule the Day. For as to the Dates of thoſe their reſpective Offices, we find no difference: Yet the *Theoriſt* declares, That the preſence of the Moon, and conſequently her rule then, was not ſo needful, becauſe *there were no long Winter-nights.* Whereas the Moon was no more made to ſhine only in *long* Winter-nights, than the Sun was to ſhine only in *long* Summer-days. And which is more, as there were no long Winter-nights then, ſo there were no ſhort Summer ones neither. So that ſet but the one againſt the other, and the preſence of the Moon may ſeem to have been as needful then, in regard of the length of Nights, as it is now.

Upon the whole matter therefore there are no good grounds for this piece of Scepticiſm. And to what has been ſaid concerning it, we need add but this, Whereas it is argued, that there might be no Moon, upon the account that there were no long Winter-nights, nor great Pool of the Sea to move or govern: we being aſſured that there was a Moon, may much better invert the reaſon, and retorting the force of the Argument, conclude that there muſt be long Winter-nights, and the great Pool of the Sea; becauſe that Planet was preſent to rule the one, and alſo to move or govern the other.

Though poſſibly the ſhutting her out of our neighbourhood, might be warily done, and with proſpect of her malignant influence in the caſe before us, namely, That ſhe might not incommode or hinder the rearing of the Earth, upon the Waters of the Chaos. For truly had ſhe been ſo near a Neighbour at firſt, as ſhe is now; ſhe might have been an injurious one as to that Affair. She might have kept thoſe Waters in ſuch Motions, as would have

have dissipated their Oily Covering; and so have put by the Primitive Earth, by marring the Basis whereon it should have stood.

Yet when all is said, I would have this Exception lookt upon as propounded in way of *Quary*, Whether the unsettledness of the Chaotic Waters, would not have hindred the Production of the first Earth? rather than as a positively assertory Objection, as if it must necessarily have done it.

4. And here I cannot but remark the exceeding *precariousness* of the *Theorist*'s Hypothesis, in reference to the Chaos, and the Formation of the Earth out of it. For that that Mass, which consisted *of*, and was then *first* dissolved *into* the *simplest* elementary Bodies in the World; should cast forth one Body (I mean Liquor) which in its purest nanatural state, could contain so much Oiliness in it. That this Oily matter should rise just when it did, so as to be fit to receive the Earthly Particles in their fall out of the Air; whereas had they come down sooner, they had been drowned in the Water. That this Oiliness should be of just such a quantity as was sufficient for use; just enough, that is, to mix with those Particles, and to make them into a good Soil: whereas if it had been more, it would have overflowed them, and made the Earth useless as a greazy clod; if less, it would not have imbib'd them, but they must have lien loose above, in a fine and dry powder, that would have rendred the Earth barren as an heap of Dust. That the Waters also should be of a due Proportion; just sufficient, that is, to make a temporary Deluge; and then to retire into the Deep, and make a durable Sea: whereas had there been much less, the Earth upon its

its Difruption, could not have been drowned; and had there been much more, it muft have been quite fwallowed up for ever. That all thefe things fhould be thus, is altogether precarious, and not to be admitted but upon better evidence, than on their behalf is given in. For here any one will be of the *Theorift*'s Judgment, as he has declared it. (*a*) That *things of moment* (fuch as he treats of) are not to ftand upon weak and tottering, *dubious and conjectural* Grounds; but to be *founded upon* SOME CLEAR AND INVINCIBLE EVIDENCE. But then he who talks at this rate, ought, when he writes of fuch momentous things; to make them out very clearly and evidently. Elfe (by what he fays more in the fame Paragraph) he proclaims himfelf guilty of a rafh attempt; even of tampering where he ought not to meddle; and of ftriving to enter at that Door, where GOD and Nature have both agreed to fhut him out. For did they think good to let him in, it fhould be by fuch a way as is *certain* (he tells us) and wherein he fhould walk with the aforefaid *evidence* on his fide. Now this, I fay, being his declared Judgment; the *Phænomena*'s above-mentioned, fhould have been more fully explained and made out; and alfo more throughly confirmed and made good.

(*a*) *Ego quidem in eâ fum fententiâ, fi in harum rerum, de quibus agitur, cognitionem, aut aliarum quarumcunque, quæ momenti funt, vifum fuerit Deo vel naturæ ut pateret hominibus ratio perveniendi, ratio illa certa eft, & in aliquâ clarâ & invictâ evidentiâ fundata: non conjecturalis, varia, & dubia, &c.* Lat. Theor. *pag.* 5.

5. But befides thofe, there is another behind, which if lookt into, will not only be found as *Precarious* as any of the reft; but alfo *Unphilofophical.* And that is, *The defcent of the Terreftrial Particles out of the Air,* which conftituted the Prædiluvian Earth.

Earth. For of those Particles the *Theory* will have *that* Earth to be made. Which were a πείψημα, or kind of excrementitious Sediment, that the Body of the Air threw out, when it purified it self. But that such a prodigious quantity of gross and feculent substance, should then lodge in that part of the Chaos; which was so light and volatile at the same time, as to * *mount above* other Bodies, and also *keep it self upon the wing*, *and play in open places*; might justly be questioned. For if such a vast deal of drossy stuff, were mixed with the Aereal matter; then whatever natural disposition (through levity) it might have, to mount up; *that*, one would think, should have so pinioned its Wings, as to have kept it down, at least from rising very high; and have been so heavy a clog upon it, as to have spoiled its playing in open places; at least its playing up so far as the Moon. Yet that the *Theory* allows it to have done so, is evident. For it supposeth them to † have *showred down not only from the middle Regions, but from the whole capacity or extent of those vast spaces betwixt the Moon and us.* A supposition that is not only precarious, but also seems (I say) to be somewhat *Unphilosophical.* For though upon the *Theory*'s account, it was necessary these Particles should fill such vast spaces; that so the Air might be able to contain *enough* of *them*, and also have *room* enough wherein to move, and by motion to purge it self, and cast them out: yet how will the *Phænomenon* fall in with a smooth Philosophic Explication? For in short, either the Bounds of the Chaos, and the Sphære of its gravity (as I may call it) did extend as high as the Moon, or they did not. If they did *not*, how came these Particles there? Especially in such plenty, as to descend from

* Theor. pag. 54.

† Theor. pag. 60.

from thence in ſhowers? Yea, how could they come down at all? Let Philoſophy make it out. In caſe the Bounds of the Chaos and the Sphere of its gravity, *did* reach ſo high as the Moon; then why did not *ſhe* come tumbling down with thoſe Particles? or rather ſooner than they, as being much heavier? Let Philoſophy give an account of that. For I think we have proved ſhe was then in our neighbourhood: though it ſeems there might be more reaſon for that Doubt, than we were at firſt aware of.

6. And as this Aſſertion is not very conſiſtent with Philoſophy, in it ſelf; ſo in the Conſequence of it, it is *againſt Scripture.* *That* aſſures us, That Light was the Product of the *firſt day.* And as it was made then, ſo it was made *viſible* in theſe inferiour Regions. But this could not be, in caſe the Earth were formed according to *The Theory*; the Air would have been ſo full of terreſtrial Dregs. For it then contained enough of ſuch Dregs, to compoſe an Earthly Orb, of above one and twenty Thouſand Miles in Perimeter; and of a depth or thickneſs we know not how great. And ſuch unſpeakable meaſures of Earth in the Air, muſt needs fill it with darkneſs; yea, with ſuch a ſpiſſitude and opacity, as would utterly have ſpoiled the Pellucidneſs of it, for a conſiderable height above the Chaos at leaſt. For the coarſeſt and heavieſt of the floating Particles, ſetling continually towards the Chaos; and the nearer they approached it, drawing ſtill into a narrower compaſs (by reaſon the ſpaces out of which they deſcended, were much larger than thoſe into which they gathered) the mighty throng of them (they being crowded toge-

M 2 ther

ther as close as their gravity could squeez them in their fall) would have made a Ring of such darkness about the Chaos, as would have been like to that which once plagued *Egypt*. It would have been palpable, that is, as containing a kind of tangible thickness and clamminess in it.

Yet in the *first day*, upon GOD's most powerful *fiat* given, there was light, *Gen.* 1.3. Which plainly argues, That the Body of the Air, could not then be of so foul a Constitution. If it had, though GOD, when he pronounced, *Let there be light*, had made the Sun (which he did not) and made him much brighter than he is; he could not have illightned these lower Regions, as being not only clouded and covered, but even stuffed (as it were) with an impenetrable density, or kind of material darkness, so far as the aforesaid Ring or Circle about the Chaos reached. But then how much less could that Light have done it, which was preexistent to the Sun, and was no more than a faint glimmering, in compare with his Glory. Yet on the first day, I say, there was Light in the Chaotic World, even on the very Waters of the Chaos. For when GOD said, *Let there be light*; where can that Light be thought to have shined more especially, than where he said before there was darkness? And where was darkness said to be before, but *upon the face of the deep*? Gen. 1. 2. And therefore Light must be shot down thither, in obedience to the Divine Command.

But then here again this *Hypothesis* seems to be unwarrantable, as grating too much upon Holy Scripture. For whereas that certifies, That there was Light on the first day, and upon the *superficies* of the Abyss (as the Context intimates) this *Hypothesis*

pothesis puts nature into such a condition, as made it impossible it should be so, and positively avers, That it was quite contrary. For it tells us, * *The Air was as yet thick, gross, and dark.* And when was it thus? Why, most certainly after the first day was past. For it was after that the immense Aereal Mass had had time to purifie it self in a great measure; as appears by what follows: *There being abundance of little Particles swimming in it still, after the grossest were sunk down.* And if the Air were thick and dark *then*, after the grossest Particles were sunk down: what was it *before*, when they were but sinking? And therefore as the first darkness, at the World's Formation, is acknowledged to proceed, † *ex ipsius Aeris impuritate & perturbatione*; *from the impurity and roil of the Air*: so the *Theory* calls it by the name of, *Tenebræ diuturnæ*; *lasting darkness.*

* Theor. pag. 57.

† Theor. pag. 243.

CHAP.

CHAP. IV.

1. A *Third Exception* against the Formation of the Earth, *the Fire at the Center of it.* 2. The *Theory faulty* in not setting forth the *Beginning* of the Chaos, which was necessary to be done. 3. *Such* a Chaos was not *Created.* 4. Nor yet produced in *Des-Cartes* his *way.* 5. And therefore that *Central Fire* seems a thing *unreasonable.* 6. That the *Chaos* was produced in the *Cartesian* way, not to be *allowed* by the *Theory.* 7. The Word, ברא, also insinuates the *contrary.* 8. The *Septuagint* cleared in *one passage.* 9. The Story of the Creation not to be restrained to the *Terrestrial World.*

1. THAT the Earth is not the solidest of the Planets, may well be inferred from its nearness to the Sun. And therefore we see *Mars* a less Planet by much, advanced above the Earth, upon the account of his solidity. And for the same reason, he may be of such a rutilant or fiery colour as he is; which Complexion (among the *Hebrews*) gives him the name of מאדים, *the red Planet.* But though that degree of Proximity, which the Earth holds to the Sun, shows her to be of a looser substance, of a more porous, and less solid nature; yet it cannot presently be improved into an Argument, of her having a great quantity of Fire at her Center. This the *Theorist* admits of as a thing

‖ Page 64. ‖ *very reasonable*; that there is a Fire at the Center of the Earth, as there is a Yolk in the middle of an Egg. But how can it be so reasonable according to his

his Hypothesis? For, according to *that*, the Earth was formed out of a Chaos, as we have heard. And that Chaos was nothing but *a fluid* Mass consisting of *earthly* Principles, as is intimated in these words, *By the Chaos I understand the matter of the Earth and Heavens, without form or order: reduced into a fluid Mass, wherein are the materials and ingredients of all Bodies.——Suppose then the Elements, Air, Water, and Earth, which are the principles of all terrestrial Bodies, mingled without any order*, &c. Now when the Chaos was a confused Mass, in its principles so wholly *terrestrial*, and in its constitution so wholly *fluid*; it is so far from being *very reasonable*, to allow a Fire at the Center of it, (and if there were not a Fire at the Center of that, how could there be one at the Center of the Earth?) that it would rather be very absurd to do it.

For so, in the First place, very contrary and discordant natures, must have been tied to dwell together in the closest cohabitation, or a perfect contiguity. In which state of conjunction o. immediate vicinity; how could they have subsisted, without preying upon, and destroying one another? Either the Fire would have dissipated the ambient fluid Bodies that were near it; or else those fluid humid Bodies, would have suffocated and extinguish'd the Fire they inclosed. Or if they could have dwelt together peaceably for a while, and not have invaded one another. Yet

Secondly, When the Chaos began to separate, and the grossest parts of it to sink down, those that subsided first (it being a fluid Mass.) must have met at the very Center of it, and the rest as they followed, would have gathered close about *them*; and so constituting a central Globe of Earth, solid throughout

throughout, would have left no hollow ſpace within it, for a receptacle of Fire.

Or Laſtly, If there had been room left for Fire at the Center of the Chaos, yet how ſhould Fire have conveyed it ſelf into that place of reception, or by any means have come to dwell there ?

2. To make this out, it was neceſſary that the *Beginning* of the Chaos, or the *way* of its entring into the World, ſhould have been declared by the *Theory*. But it is not done: which ſeems to be a kind of flaw in the *Hypotheſis*. It takes no notice of the cauſe of its Origin, nor of the manner of its Production; whereby this difficulty might have been prevented or cleared up. And truly the way or manner of its riſe or emergency into being, is neceſſary to be known for the explaining of *other* difficulties, as well as this. For upon it depends the ſolution of ſeveral Phænomena's, and very material ones. I name but one, *The Magnetiſm of the Earth*, as to the influence it has upon the *Index nauticus*, or *Needle of the Mariner's Compaſs*; the pointing or Direction of which, is not ſo curious and ſurprizing; but it is as uſeful in the affairs of human Life. But then if the *Theoriſt*, (by ſetting his Chaos, which came from we know not whence, in the room of an Earth of a Planetary Origin, ſunk down from its lucid or Sidereous ſtate) takes away the ſuppoſed cauſes of this notable effect; it will be incumbent on him to aſſign others, from whence it may be derived.

In caſe it be objected, that the Phænomenon alledged is not ſatisfactorily accounted for in the *Carteſian* way neither; foraſmuch as it ſtumbles in the formation of the *Striatæ Particles*, the main inſtrument of of the work; and that *Des-Cartes* himſelf dares not truſt his own Hypotheſis, but profeſſes the Earth to have been otherwiſe produced than *that* determines (as ſhall be noted by and by) then I anſwer: As this is really nothing to us, ſo it will not excuſe the *Theorist* in the leaſt, from clearing up the thing, according to thoſe meaſures he hath taken by himſelf. It only ſhows, that the *French* Philoſophy (of ſo great fame) is too ſhort to fathom the deeps of Nature, and by no means quick-ſighted enough to ſee to the bottom of her profound Myſteries. Though that Philoſophy may grow up apace, to ſo happy a perfection, as to be able to make a more full diſcovery of ſuch ſecrets; muſt needs be the deſire of wiſe and good Men.

And ſo we return to the Enquiry we were upon, *viz.* How Fire ſhould come to the Center of the Earth. Which is a Problem the more intricate and perplexed, in regard *The Theory* takes no notice of the beginning of the Chaos. It tells indeed that there was a Chaos, and what kind of one it was; but it gives no manner of account how it came into Being. As to *that* the Reader is left at a loſs, and has nothing to guide him but his own Conjectures. I ſhall gueſs therefore as fairly at the thing as I can; And to me it ſeems probable, that this Chaos ſhould be produced one of theſe Two ways; either by *Creation*, or by *Des-Cartes* his way *for generating Planets*. Though it will not be over eaſie to make out, That it came into exiſtence by *either* of them.

3. For firſt, to affirm that it dropt directly out of the hand of Omnipotence, in way of *Creation*; is more than we find warranted. Yea, we are

taught something, and that from Heaven, which is
* Gen. 1. 1. very different from it; Namely, That * *in the beginning GOD created the* EARTH. And if it was an *Earth* that he created in the *beginning*, it could not be a *Chaos*; I mean *such* a Chaos as the *Theory* makes it: for that was no *Earth*, nor had it any specific or distinct *Earth* in it, as being † *without distinction of Elements.* It is said indeed, *Gen.* 1. 2. That the Earth was, תהו ובהו, *desolation and emptiness. Inanis & vacua*, as the *Vulgar*, doubly void. That is, of its designed order and comeliness, which were to beautifie it: and of all those creatures which were to furnish it, and dwell in it. And therefore, says the *Targum* of *Jerusalem*, *it was empty* מן בני אנשא *of the Children of Men*; וריקניא מכל בעיר *and void of every brute.* And the Prophet describing a most fearful destruction to come upon his People by Wars; through which their fruitful Land was to become a Wilderness, and Men and Birds were to be driven away: tells us in the very Words of *Moses*, That the Earth was, * תהו ובהו, *desolation and emptiness.* And in this sense I confess, the Earth (in its original imperfection and nakedness) was a *Chaos*: an incultivate and uninhabited lump, rude and confused beyond all imagination, as having neither good form nor furniture in it. But then at the same time it was *an Earth* too; and so not such a Chaos as the *Theory* speaks it. I might also note (would *that* be of weight) that the *Præfix*, ה, in הארץ (*Genesis* 1. 1.) is ה *notificationis*, (הירעה) *scientific* or demonstrative: and so it points at *this* Earth, and intimates it was *this* very same Earth at *first*, that it is *now*. The same as to substance and nature, though not as to condition and ornaments. And *this* Earth, in the state of its primitive disorder

† Theor. pag. 44.

* Jer. 4. 23.

diſorder and deſtitution, being the true *Moſaic* Chaos, created in the beginning; we have no grounds to believe, that any other beſides it was ever brought forth in way of *Creation*. But we have good grounds to believe that no other was ſo produced; inaſmuch as to aſſert it, would be to ſet up Phantaſie againſt *Moſes*'s authority; and to bring preſumptuous conceit, into competition with Scripture.

But grant the Chaos to have been ſuch as it is ſuppoſed to be, and that it entred into the World at the door of Creation: Yet here will be nothing to make it reaſonable, *very reaſonable* to admit Fire at its Center. For if there was a central Fire in it, it muſt either be placed there *ſupernaturally*, by the immediate power of the ALMIGHTY; and we have no reaſon to admit it upon that ſcore, becauſe we are no where informed of it. Or elſe it muſt be generated there in a *Natural* way: and to admit *that*, would be againſt reaſon too. For how could a vaſt quantity of Flame, be bred in the Bowels of an Earthy Maſs, conſiſting of the Principles of all terreſtrial Bodies. And whoever ſhall peruſe the firſt half of the *Fifth* Chapter of the *Engliſh Theory*, will ſoon be ſatisfied that the Chaos *could* conſiſt of no (*a*) *other* but terreſtrial Principles. For there it appears that it was reſolved into nothing but Earth, Air, Water, and an unctious ſubſtance; and ſo could be made up of nothing elſe. But Fire is quite another thing; and as different from thoſe Elements, as motion is from reſt; or the moſt Celeſtial, from the moſt Terreſtrial Matter: and ſo in a courſe of Nature could not poſſibly iſſue from them, and ſettle it ſelf in the midſt of them.

(*a*) The *Schemes* of the Chaos ſhow it *terreſtrial* throughout. *Theor.* pag. 54, 55, 56, 57. The *Earth* alſo formed out of it, is repreſented *without Fire* at the Center, pag. 58.

4. We will pass therefore to the *Second Conjecture*, whither indeed the Notion of central Fire in the Earth, does most directly lead us :, and that is *Des-Cartes*'s Method, by which he supposeth Planets to be formed. And according to *that* the Earth was one of those *fourteen*, or *fifteen* Stars, which once shined gloriously in their respective Heavens hereabouts. But being all overgrown and incrusted with *Maculaes*, except one, and losing their native strength and light ; were swallowed up by the *Vortex* of the surviving Luminary the Sun: and so move round about him as so many *Satellites* or Waiters of his, to this day. Though some of these Planets also, that is, *Secundary* ones, are at the same time carried about others of them. As the Moon, about the Earth ; the four *Medicean* ones, about *Jupiter* ; and *Saturn*'s three *Assecla* or Pages (according to *Cassinus*) about him.

And here there may seem to be a plausible account given, of the declared Central constitution of the Earth, or of a Region of Fire at the heart of it. For it having been all Flame heretofore ; till it was overspread with *Maculaes* (boiled out of it self, and gendred first into a kind of foam or scum ; and afterwards into an harder substance) it could not but retain much fiery matter in its Central Parts. And thus this Fire would be sufficiently protected too, against dissipation and danger of Extinction, from the moist and lumpish Chaos which surrounded it, and at the time of its separation would have lain heavy upon it. For its Coat of *Maculaes* worn next it, being nealed by furious heat, and made into a strong arched Vault ; there the inclosed Element might have been secure (as in a mighty Granado-shell)

Granado-ſhell) never to be annoyed by any manner of violence.

But neither by *this way*, as quaint as it is, could the Chaos ſtep forth into being. For though it be a ſpruce and gay invention; a contrivance rarely ingenious, and prettily coherent; and withal ſo laudably inſtrumental to the trim ſolution of ſundry difficulties, that ſome are ready to think 'tis pity but it ſhould be real: yet the very firſt daſh of *Moſes*'s Pen, gives the Philoſophic Bubble ſuch a ſhrewd prick, as flats it into Vanity and Romance. *In the beginning GOD created the Heaven and the Earth.* So that famous Man told an illiterate People, as a faithful Secretary of the MOST HIGH; with intention fully to inſtruct them as to the Origin of the World, ſo far as comported with his Majeſtick Office and brevity. And if GOD in the beginning, at the very firſt, *created* the Earth, and created it *an Earth*; how could it before that, be a *Chaos, ſuch* a Chaos, as it is repreſented to have been? and how could it poſſibly riſe into ſuch a Chaos, out of a Sun or fixt Star? And if GOD created the *Heavens* at the *ſame time* when he created the Earth (as *Moſes* affirms) for *both*, he ſays, were created *in the beginning*) where could it have place to act the part of a fixed Luminary, before it became a Planet? But therefore to take off this, and the like Arguments, the Story of the Creation is ſuppoſed to relate to the *Earthly World* only. How *truly* we ſhall a little conſider in the Cloſe of this Chapter.

In the mean time, to go on with what we have in hand; the illuſtrious *Des-Cartes* is on our ſide. He openly profeſſes (as was noted * above) that he did not think the Earth was made of a Star, according

* Chap. 2. § 4.

according to his Principles; but was brought forth by Creation. And he judged thus, for the same reason, I am now urging. Because, † *hoc fides Christiana nos docet*; *the Christian Faith teacheth us as much.* So that he who shall teach otherwise, must (in the opinion of that renowned Philosopher) broach a Doctrine against Divine Revelation. And therefore what has been said that way, I hope will relish the better, as falling in with the sentiments of so exceeding worthy and judicious a Person. And herein he acted like a true and noble Christian Philosopher indeed; in that he made his *Hypothesis* stoop to the Religion of Heaven, and would retrench his Principles, rather than they should run counter to the sacred Oracles. Yea, the great Man goes farther, and adds, *hocque etiam ratio naturalis plane persuadet*; *and this also* (that the World was created with all its perfection, so that there was in it a Sun, and an Earth, &c.) *natural reason does plainly perswade.* (a) *For if we attend to the mighty power of* GOD, *we cannot think that he ever made any thing that was not compleat in all points.* And therefore he said before; (b) *And likewise in the Earth there were not only the seeds of Plants, but Plants themselves; nor were* Adam *and* Eve *brought forth Infants, but made adult persons.* And when it is a thing not only worthy of GOD, to make Creatures perfect at first, but *natural reason perswades* that he actually did so: we must either conclude that the Earth was made so (as *Des-Cartes* does) or else in our judging otherwise, vary from, or go against the dictate of common reason, as well as Scripture. So

† *Princip. part.* 3 *Art.* 45.

(a) *Attendendo enim ad immensam Dei potentiam, non possumus existimare illum unquam quicquam fecisse, quod non omnibus suis numeris fuerit absolutum.* Ubi supra.

(b) *Ac etiam in terra non tantum fuerint semina plantarum, sed ipsae plantae; nec* Adam & Eva *nati sint infantes, sed facti sint homines adulti.* Ibid.

So that if the Opinion, the professeed and openly avowed Opinion, of the most eminent Christian Philosopher; yea, of the admired Author of the new Philosophy (the fittest person amongst Philosophers to judge in the Case) will cast the Scales for us; we have it on our side, that the Earth was not produced in his way; or according to his *Hypothesis*.

5. But then the premisses considered, to admit there is a Fire at the Center of the Earth; is so far from being *very reasonable*, as the *Theory* holds; that according to the fairest measures or accounts of things, which Philosophy has given to the World as yet; it rather appears to be very *unreasonable*. For however *Des-Cartes*'s Principles lean that way, and countenance the Phænomenon; yet he himself, we see, not only doubts of his own *Hypothesis* as to the Earth's formation; but has publickly declared that they who sail by his Compass, must swim against the stream of *Natural Reason*.

6. And truly should the *Theory* allow this Central Fire in the Earth upon the account of its being produced in *Des-Cartes*'s way; it would quite overthrow its *own* Hypothesis, by complying with *his*. For then the Earth could never have been of an Oval Figure. Nor could it have been without Mountains, and without a Sea. But its motion of inclination would have been from the first, because its *Axis* would not have been perpendicular to the plain of the Ecliptic. And so its Equinoctial position (to name no more *Essentials* of the *Theory*) would have been impossible. And whereas (by the way) the present site of the Earth (which might seem more convenient, were it placed so, as that its Annual and Diurnal motions might be both performed

med on parallel *Axes*) is made by * *Des-Cartes*, to depend upon the influence of the *Striate Particles*; and both the formation and motion of them are shewed by a † learned Philosophic Pen, not to fall in with Mechanical Laws: this will be no check or difficulty upon us. For first, *Des-Cartes* might in that (as he has done in some other things) keep too strictly to the Laws of Matter and Motion; it being necessary in the works of Nature very often to acknowledge the hand of Providence. Or else secondly, if there should be no such particles in being, and nothing of their power to hold the *Axis* of the Earth, in a parallelism to that of the World's *Æquator*; this would be but an advantage on our side. For how can the Earth have a Fire at its Center as being produced in conformity to the *Cartesian* Principles, when, according to those Principles, it could not be produced at all? For put by the Formation of these Particles, and (according to that Philosophy) there could be no Planets, and so no Earth: the matter of the third Element being not to be made without those particles.

* Princip. part. 3. Art. 155.

† Dr. More Epist. ad V. C.

7. Were I dispos'd to follow the *Rabbies*, I might here go a little farther still. I might venture to lay hold of the Word, ברא *he created*; and make it do service upon their authority. For some of them bear us in hand, that it denotes, *Creation*, in a rigorous sense; that is, the making of a thing out of nothing. Agreeable to which is the holy Writer's notion of Creation where he says, * that *things which are seen, were not made of things that do appear*. Meaning (as we read elsewhere) that they were made, ἐξ οὐκ ὄντων, † *out of nothing*. Which apply to the making of the Earth (as we very well may, ברא, being

* Heb. 11. 3.

† 2 Macc. 7. 28.

ing the SPIRIT's word concerning it) and it could not possibly be made out of a Sun, or Star, as the new Philosophy would have it. For then (say those Doctors) a more proper word should have express'd its production, viz, עשה, which imports the making of a thing out of præexistent matter. Some slight ground for this seems to be laid in Scripture; and that in *Moses*'s *Cosmopæia* too. For it is said, *Gen.* 2. 3. That *GOD rested from all his work*, אשר ברא לעשות, *which he Created to make.* Where unless we allow a distinct signification, to the two words implying that he made some things out of *nothing*, and others out of *prepared matter*: we must charge the HOLY GHOST with indecent Tautology. Though if we consider again that the word ברא, is used promiscuously to express GOD's making of things *ex præjacente materia, out of extant matter*, as well as out of *nothing*, and that in the very story of the Creation: we may well suspect that there was too much niceness in *the Masters*, rather than such respective significations; to be strictly and continually appropriated to the words. Only as many as did thus criticize, have thereby fairly given their suffrage, for that truth which we contend for. That when GOD created the Earth (according to *Moses*'s narrative) he educ'd it directly out of *nothing*. And so it cannot have a fire at the Center of it, because it issu'd forth into being; in *Des-Cartes*'s way.

8. Being unawares fallen upon that expression אשר ברא לעשות *which he created to make*: in reverence to the *Seventy*; I cannot but take one short step out of the way, to vindicate their translation of it. They render it ὧν ἤρξατο ποιῆσαι, *which he began to make.* As if the work of Creation had not then been consummate. But that could not be their meaning. For

whereas we read in the beginning of the Chapter; *the Heavens and the Earth, and all the Host of them were finished*: ויכלו, they rendred by, καὶ συνετελέσθησαν, *and they were compleated.* They could only mean therefore, that *GOD rested from all his work*, which at any time, in the six precedent days, *He had BEGUN to make.* And so their sense is sound and true, though they keep not close to the literal strictness of the Original. And that they thought the Creation was wholly perfected before the seventh day; is apparent from that liberty they took, in translating the beginning of the second verse of the same Chapter; which perhaps is more culpable. For whereas the *Hebrew* says, GOD had ended his work, ביום השביעי *on the seventh day*: departing, quite from the proper signification of the word, they render it ἐν τῇ ἡμέρᾳ τῇ ἕκτῃ, *on the sixth day.* As if they feared they should offend (by stretching the work of Creation too far) in case they had turned it, *GOD had ended his work on the* seventh day. Here was more than *abundans cautela*, too much caution used. Especially if *Aben Ezra*'s Maxime be authentic, כלוי מעשה איבו מעשה * *The* finishing of a work or consummation of it, is not the work it self. I have noted this the rather, as containing in it a full consent, with what has been said touching the Creation's being perfected in Six days. For it makes it evident that the LXX. Interpreters were throughly perswaded of this Truth. And not only so, but forward to assert, and resolved to maintain it, even to an over acted care and blameable Scrupulosity.

* Ad Exod. cap. 12. v. 16.

9. And here it will not be amiss, to reflect a little upon one notion of the *Theory*'s; which countenances the late production of the Earth, or its rising long

long after the World was made (perhaps out of a Sun or Star, as the Scheme in the *English Theory*, before the Title Page, plainly insinuates) And that is, *The limiting of* Moses's *Story of the Creation, to this lower World*: to the Earth, that is to say, and the Aereal Heavens, and such things as were formed out of the Chaos. Thus in one place it confines it. * *First that must be noted, that* Moses *did not describe the first production of matter, and the rise of the universal World, but the formation of our World, that is of our Earth, and our Heaven, out of their Chaos.* And presently after; † *But the Subject of* Moses's *Genesis is the Chaos, and that most confus'd and Earthly; and the things made out of this Chaos, and related to it as a center; those properly belong to the* Mosaic *World.* And by and by, * *We may not surmise therefore that when we and our World was made, entire nature must needs be made at the same time.* And then again, † *Certain* it is-------*that* Moses's *World does not comprehend all the Regions of the Universe, nor all the orders of things, but those parts of Nature which could be made of the Earthly Chaos.*

* *Illud primò notandum est, non id agere* Mosen, *ut primam materiæ productionem, atque Universi Mundi ortum describeret, sed mundi nostri, scilicet telluris nostræ, & Cœli nostri è suo Chao formationem.* Theor. lib. 2. cap. 8.

† *Subjectum autem Geneseos* Mosaicæ *est Chaos, & confusissimum & terrestre; & quæ ex hoc Chao educta sunt, & ad illud tanquam centrum referuntur, ea propriè spectant ad mundum* Mosaicum. ib.

* *Ne putemus itaque nobis nascentibus & mundo nostro, necesse esse ut tota natura eodem tempore nasceretur.* ib.

† *Pro certo & explorato habeatur* ——— *Mundum* Mosaicum *non omnes Universi regiones, neque omnes rerum ordines complecti, sed illas naturæ partes quæ è Chao terrestri educi potuerunt.* ib.

But then (to say nothing of *Light*, or the *Vehicle* of it, neither of which were made out of the Chaos) let me ask; What did GOD mean, when he said, יהי מארת * LET THERE BE LIGHTS? I do not ask, what those Lights were, that's evident enough. Nor where they were placed: for they were far above the Aiery Heaven, and so in the sense of the

* Gen. 1. 14.

Theory, could not belong to the Earthly World. But the queſtion is, What ALMIGHTY GOD intended by, LET THERE BE LIGHTS. The *Theory* * hints the meaning and effect thereof to be no more, than that thoſe heavenly parts of the Univerſe, were then firſt made *conſpicuous*, or began to illighten the Earth: and declares it *demonſtrable*, That *Moſes* is ſo to be underſtood, as he has limited him.

* *Dici poſſent tum naſci & oriri ex partes cœleſtes Univerſi, cùm primùm conſpicuæ erant, atque diſſipatâ caligine Chaos & nigri aeris, eminus ſe oſtentabant terris, paulatim emicantes è tenebris, quaſi ab iiſdem & eodem Chao, enatæ fuiſſent. Neque aliter Coſmogoniam* Moſaicam *intelligendam eſſe, ſi opus eſſet, me demonſtrare poſſe exiſtimo.* lib.2.cap 7.

But then I muſt continue the enquiry, What does, יהי, LET THERE BE, ſignify in *other* places of the ſame Chapter; where it occurrs ſo often? why, it infallibly implys the *production* of thoſe things, to which it does reſpectively relate. It imports God's commanding or willing their exiſtence; and their immediate emergency into being, in obedience to his powerful Will or Mandate. This is obvious even to ſlighteſt notice. Thus, when GOD ſaid, *Let there be light*: it follows immediately, *and there was Light.* When GOD ſaid, *Let there be a firmament*: it follows, *and GOD made a Firmament.* When GOD ſaid, *Let the waters be gathered together into one place* (that ſo there might be dry land, and Seas:) it follows, *It was ſo.* When GOD ſaid, *Let the Earth bring forth Graſs, &c.* it follows, *and the Earth brought forth Graſs, &c.* When GOD ſaid, *Let the Waters bring forth abundantly:* it follows, *the Waters brought forth abundantly.* When God ſaid, *Let the Earth bring forth the living creature after his kind, and Cattel, &c.* it follows, *And GOD made the Beaſt of the Earth after his kind, and Cattel, &c.* And when the Divine and Omnipotent *Fiat* did all-along carry ſuch energy with it, as thus to produce other things;

things; as in the ſeries of the Story: can it in reaſon be thought to do leſs, when GOD pronounced, LET THERE BE LIGHTS? To make this one *Fiat*, differ in ſenſe from the reſt; would be to depart from the Rules of a juſt Expoſition. Yet unleſs we force ſuch a difference into it, it muſt ſignify more than the bare *appearance* of lights upon the clearing up of the Chaos and the Sky: that is, it muſt ſignify thoſe lights were juſt then *created*.

And this is farther evident thus; in that GOD takes notice, expreſs notice of the *uſe* of theſe Luminaries, and therein particularly provides for the *conſpicuity* and *radiancy* of them. † *Let them be FOR lights in the firmament of the Heaven, to GIVE LIGHT UPON THE EARTH.* So that when he ſaid, LET THERE BE LIGHTS, if he did not mean more than their becoming *conſpicuous* and *ſhining out upon* the Earth; the two expreſſions muſt be perfectly *tautological*. And yet if he intended any thing elſe, what could it be but their *Creation* at that time? Eſpecially when it follows hereupon, *And GOD made two* * *great lights*, and *the Stars alſo*. And therefore that the work of Creation which *Moſes* treats of, reaches farther than what belongs to the Earthly World, and reſulted from the Chaos; is not to be doubted. For he does not only mention the *making* of the Lights in the Firmament (things as different from the terreſtrial World, as they are diſtant from the ſame) but deſcribes them as fully, in relation to their *uſes* and ends; and ſo ſeems to handle them as profeſſedly, as any piece of the lower Creation whatever.

† Gen. 1. 15.

* The Moon is really a great light to the Earth, though the light ſhe tranſmits thither, ſhe borrowed of the Sun.

In caſe it be objected, that the *Stars* give little light upon the Earth, which is a thing *Moſes* aſcribes to the Luminaries in Heaven; I anſwer, If they ſerved not ſo eminently to *that* uſe, yet to the *other*

he

he mentions, they were very serviceable and indispensibly necessary. For how could time have been measured out and divided into *Years* and *Months*, (as it was in the First World) without their help? especially if there were no Moon.

And so I demand in the *Second place*; What does *Moses* mean, by *the Host of the Heavens being finished*?
† Gen 2. 1. *Thus* † *THE HEAVENS were FINISHED, and all the HOST of them.* If he meant only the Host of the Heavens belonging to the Earth; what was the *Host* of those Heavens? As for the Air, it helped to constitute them, to make the very Heavens themselves. As for Clouds, Rain, Hail, Snow, and the like Meteors; there could be none, says *the Theory*. As for the Moon, it might not then be in the Earth's Neighbourhood. As for that watry exhalation which abounded in the aereal Heaven, it was but one single thing; and so answers not the import of the Word, HOST, it being of a plural signification. And what other Host should belong to these Heavens, except the Fowls? but then though in Scripture they be called, עוף השמים; and
* Dan. 2. 38. in the *Chaldee* * עוף שמיא; and by the *Septuagint* and in the *New Testament*, πετεινὰ οὐρανοῦ, *the Fowls of Heaven:* yet I do not remember that they are any where called, צבא השמים, *the HOST of Heaven.* That phrase in Holy Writ, does *usually* (I think) *continually* referr, either to the *Angels*, or else to the *Lights* of Heaven. And of the *latter* of those at least, it must here be understood. But then none of these Luminaries being formed out of the Chaos; and all of them but one, placed in remote or superiour Heavens: hence it is evident, that the Story of the Creation, is not to be restrained to the Terrestrial World. For that *Moses* did not only speak of them, but

but of their being *created* then; is manifest from the words before us. *The HEAVENS and the EARTH* were *FINISHED*, and all the *HOST of THEM*. Where, if by the *Earth* and *its Host*, being *FINISHED*; we are to understand their being CREATED at that time, as we certainly must: then are we bound to understand, that the Heavens and *their Host*, were so too; because the same thing is equally predicated of *both*. It may be worth the while also, to remark that Passage in the 148. *Psalm*. Where the inspired Man desiring that GOD might be glorified by means of the Celestial Luminaries; crys out, *Praise ye him, sun and moon: praise him all ye stars.*——— *For he commanded, and they were created.* Whence it is evident, that when GOD commanded, *Let there be lights*; this was not a command whereby they *appeared* only, but whereby they were *created*: and the *Moon* with the rest was *then* commanded into being.

I might also make a *Third* demand, What is meant by the נשמת חיים * *breath of life*, which GOD breathed into Man? No less than his *very Soul*. So says † *Buxtorf* (and * others) *the* Hebrews *by* נשמה *understand the rational and immortal Soul, and therefore they swear by it.* And when GOD *created man*, did he not create this Soul of his? And so did not the work of Creation, which *Moses* writes of, comprehend more than those parts of Nature, which were made of the Earthly Chaos.

* Gen. 2. 7.

† *Lexic. in* נשם.

* *Usurpatur de homine tantùm, & animam hujus ratione præditam denotat.* Schind. *Lexic. Pentag.* pag. 1177.

Dicitur propriè de anima hominis immortali, quam Deus in illum insufflavit. Eithn. *Lyr. Prophet. in Psal.* 18. v. 16.

It may be not, will *Platonists* say; at least this instance is no good Proof of it. For GOD might not create the Souls of *Adam* and *Eve* just then, but

but send them down from a state of *Preexistence.* But then (not to ingage in a new Controversie) I reply in short; If the humane Souls came into their Bodies out of a state of *Preexistence*; then when they descended, they were either pure from sin; or they were not. If they were not pure, then how did GOD create *Man in his own Image?* Gen. 1. 27. Or how did he make *Adam upright?* Eccles. 7. 29. Where the *Rectitude* spoken of, must be of a moral nature; because (as the Context shows) it is opposed to moral obliquity or perverseness. If they were pure, how could the infinitely gracious BEING (whose name (and so his nature) is † MERCIFUL; who * *delighteth in mercy*, and whose † *mercy is over all his works*) deal so unkindly with his own most dear and spotless creatures; as to thrust them down, or suffer them to fall, out of a state of Æthereal light and happiness, into a state of darkness and stupid *silence*, out of which (according to *Platonism*) they must come, to be incarnate, and so slide into a condition more forlorn still? Truly if the goodness and wisdom of Heaven, so decreed and ordered things, as that the Protoplast's (and so their Children's) innocent and immaculate Spirits must be betrayed or precipitated into that state of *inactivity* (which might last for millions of years or ages) and then out of that squalid condition, sink into a worse; into one full of inexpressible imperfections, miseries, and dangers, where innumerable multitudes lie under almost an inevitable necessity of falling into the torments of everlasting destruction: if this, I say, be the result of Heavens wise Councils and Decrees, *Preexistence* will give no satisfaction to understanding Men; and do as little honour to the Glorious GOD. It will

† Exod. 34. 6.
* Mic. 7. 18.
† Psal. 145. 9.

will rather be a Scandal, than a Key to Providence.

Now that the Souls of the first Pair of Mankind did preexist, it being improbable; and that they should be *ex traduce*, it being impossible: what remains, but that GOD *created* the Souls, when he made the Bodies of those Persons? And so the work of Creation, of which *Moses* treats, is so far from being limitable to the *lower* World, or indeed to the *higher material* one either; that it stretches out it self beyond them both, even to the *Spiritual* one. And *the Host of the Heavens*, just now done with, intimates as much, Expositors conclude, while they make it refer, not only to the *Lights*, but the *Angels* above.

And perhaps something of this Truth, That Angels and Humane Souls came into being at the same time that the Earth did; may be wrapt up in the Doctrine of the *Mundane Egg*. So *Orpheus*, that ancient and famous Divine amongst the Heathens, who, according to * *Athenagoras*, πεπίστευται ἀληθέστερον θεολογεῖν *is believed to Theologize more truly than the rest*: tells, how ὑπερμεγέθης ᾠὸν, *a superimmense Egg* being brought forth by *Hercules*, that is, I think, by the Divine Power; ἐκ παρατριβῆς εἰς δύο ἐῤῥάγη, *by attrition it brake into two parts*: of the upper part of which, was made Heaven; and of the lower part, the Earth. And then affirms, that Heaven being mingled with Earth, it produced both Women, Men and Gods. By which he might shadow out that the Intellectual or Spiritual World, took its beginning, with the Terrestrial one. But if he meant that Souls or Spirits sprung up out of matter, this will make the ancient Philosophy so very mean and gross, as not to be at all regarded.

* *Legat. pro Christian.*

P CHAP.

CHAP. V.

1. *The* Form *of the Earth* Excepted *against, from the* want of Rivers. 2. *Notwithstanding the* way *devised to raise them, there would have been none in* due time. 3. *Whereupon Two great Inconveniences must have ensued.* 4. No Rivers *could have been* before the Flood.

(a) Ecclus. 39. 21.

1. THE (a) *chief thing for Life is Water*, said the Son of *Sirach*. It is necessary and useful upon numberless accounts. So that *that Hypothesis* which implies the Earth was without Springs and Rivers for many hundreds of Years, may justly be rejected. And for this reason the supposed Form of the Earth cannot be maintained. For according to that, the Element of Water was fast shut up within the Exteriour Orb of the Earth; and how could it issue forth from thence, through so thick and solid a terrestrial Concretion? For that being made after the manner abovesaid, there could be no gaping chasms, nor indeed little clefts or chinks in it; whereat the imprison'd Waters might get out. Or if there had been never such plenty of lesser cracks or larger rifts in it; yet the Water being settled in that place, which was proper to its Nature, there it would have staid by the innate Law or Principle of its Gravity. Unless by Elastic Power, Protrusion, Rarefaction, or the like, it were forced thence; there it would have made its perpetual aboad, had the Earth been never so open or pervious, by reason of fissures or holes in the same.

2. But

2. But therefore *Exhalation* is here made uſe of, and as a proper Engin is ſet to do this mighty work, of fetching up Rivers from the inacceſſible Pit. The operation, in ſhort, was performed thus, The heat of the Sun raiſing plenty of Vapours, chiefly about the middle parts of the Earth, out of the ſubterraneous Deep; they finding moſt liberty and eaſieſt progreſs, toward North and South, directed their motion towards the Poles of the Earth. Where, being condenſed by the cold of thoſe Regions into Rain, they deſcended in conſtant and exuberant diſtillations. And theſe Diſtillations were the Fountains that ſupplied the firſt World with Rivers, running continually from the Polar to the Equinoctial parts of the Earth. But according to this Hydrography, I ſhall endeavour, firſt, to make it out that there could be no Rivers in due time: and ſecondly, that there could be none at all. And as for Springs, the *Hypotheſis* does not pretend to any.

Firſt, It would have kept Rivers too long, out of being. For according to that Philoſophy we have now to do with, the new made Earth was compoſed of nothing but Duſt and an Oily liquor. And it being of ſuch a Compoſition, and of a vaſt thickneſs; it muſt needs be a conſiderable time before the Sun could penetrate into the Abyſs under it, and draw up vapours from thence; if it could do it at all in ſo copious a manner.

Secondly, The Air being at firſt quite empty of Vapours, it would take a great quantity of them to make the Atmoſphere of the Earth, or to fill up that. To which add that every part of the Earth about its Æquator, being turned from the Sun, every four and twenty hours, as long as it was

 obverted

obverted to it; many of those Vapours which were lifted up by day, would fall down again by night in the same Latitude where they arose, without being dispersed to the Polar Regions. And thus the production of Rivers would have been something retarded again.

Thirdly, The surface of the Earth being endued with a wonderful feracity it must immediately put forth in an inconceiveable plenty of all sorts of Vegetables: which from luxuriant pullulations, would strangely advance by most speedy and prodigious growths. And this Superfetation of the virginal Soil, proceeding from that extraordinary fruitfulness wherewith it was originally impregnated; must farther hinder the early rise of Rivers. Not so much by consuming the matter of them, as another way. For the Earth being thick beset with the flourishing *apparatus*, or goodly Furniture of its own bringing forth (such perhaps for abundance and excellency, as never crowned the most fertil Country, or fruitful season since) though Dews or Rains fell without intermission; yet the Waters would have stuck or hung so much, amongst the rank and matted tufts of Grass, Herbs, Shrubs, *&c.* as not to have been able in a short space of time, to have gotten into Streams, and constituted Rivers of such a length, as they must have been of.

Fourthly, In case these Waters had met with no checks, but had fallen immediately into such Bodies, as would have forced their passage along in holding Currents: yet then they must have digged their own Chanels too, being sure to find none till they made them. But considering how slowly they must have crept, as having no kind of Precipices or steeper downfals to quicken them; and how glib they

they muſt have been, by gliding gently upon the fat and viſcous Glebe; and what a thick and thrummy and cloſe wrought Mantle the Earth then wore: for them to have furrowed out deep and winding paſſages in that Earth, muſt have been a good whiles work again, if fecible at all without the help of Art. For

Laſtly, It ſeems improbable that any Rivers, without the help of *that*, ſhould have been produced. The reaſon is this, The Rains deſcending at all times, and in all places alike, a round the Poles; and the whole ſurface of the Earth being more level and even, than any Plain in the World; the Waters inſtead of parting into ſtreams, would have ſpread over all the Earth at once, in a general diffuſion; as any one may find by pouring Water upon a Globe. By which overflow, the Primigenial Soil (which was a light and ſoft Mold) being ſuppled into a perfect Moor or Quagmire; muſt have continued drowned, till by reducing the Water into artificial Canels, it could have been laid dry. But when there would have been hands for this great work (GOD making Mankind but in one ſingle Pair) let them that pleaſe conſider. And they may think alſo, where Paradiſe could have been; and what ſhift poor Fowls, and Beaſts, yea, Men themſelves ſhould have made; till the Earth, like a Fen, thus under Water, could have been cut and drained.

3. Now ſo ſlow and late a Production of Rivers, would have drawn two great inconveniences after it. It would have claſhed with Scripture; and charged Providence with Prepoſterouſneſs.

First,

First, It would have clashed with Scripture. For no sooner was Man created and placed in Paradise, but presently we read, That a *River went out of* Eden *to water the garden*, *Gen.* 2. 10. But had all Rivers come into being, as the *Theory* teaches, one could never have been there so early. Nor did it go out of *Eden*, by running through it only, but it arose there, say some, and as much is signified, they would perswade us, by the word יצא which denotes its *going out* (they tell us) as a Child goes out of the Womb: and so the River must be born in *Eden*, or spring up originally there. But the Word is too commonly used, in a larger sense, both in the Sacred and Rabbinical Writings, to have any such stress laid upon it. Though most certain it is, that a River there was in *Eden*; and in order of Divine Story (and so why not in order of time?) very early: even before the Fall of *Adam*, or the Formation of *Eve*. And which is farther remarkable, it was a large River too; for it was parted into several Heads, and able to feed most considerable streams. One of which, namely, *Euphrates*, is reckoned among the biggest Rivers in the World, to this day. But had it come by derivation from the Polar Fountains, it could never have been made so soon; much less could it have been so large. And then besides, we read at the Sixth Verse of the same Chapter, that GOD *had not caused to rain upon the Earth* as yet; and so that River could not possibly proceed from Rains, that fell about the ends or Poles of it.

Though (by the way) how that Expression should countenance an *Impluvious state* before the Flood, as * the *Latin Theory* would make it, is not so clear and easie to be understood. For, if we consider, there was no Water upon that Earth, but what fell in Rain. And

* Page 206.

And in two Regions of that Earth, there were Rains continually defcending: and they feem to have been of little other ufe, than for thofe Rains to come down in. And to fay, That by the Earth *there*, was meant only *Regiones culta*, or the *inhabited Countries* of the Earth; would be an unwarranted reftriction of the Scriptures fenfe. For in the Story of the Creation, הארץ, *the Earth*, is ftill put (as we may obferve) for the *entire Globe* of the Earth, or at leaft for כל־פני הארמה, *the whole face of the ground*, as *Gen.* 2. 6. Nor may it be faid to be fpoken *ad captum vulgi*, as *to the common peoples apprehenfion*. For furely they were not fuch dull Souls, in the firft World, but (had Nature ftood in that order as the *Theory* fets it) they would have traced their Rivers to their heads, many hundreds of Years before the Deluge; and have been generally and throughly acquainted, with thofe Rains by which they were raifed. They would then have known as well, that Rivers came from Rains at the ends of the Earth; as we do now, that Gold comes from *Guinea*, or the diftant *Indies*. Yea, the want of room (they multiplying exceedingly) would have forced them to find out the rainy Regions, while they muft have fpread their Colonies to the Borders of them.

Secondly, It would tax the Providence of Heaven with Præpofteroufnefs. That is, in reference to one fort of Animals, the Fifhes. For then they muft have been brought into being, before there were fit Receptacles for them. I confefs, GOD *faid*, *Let the waters bring forth abundantly the moving creature that hath life*, *Gen.* 1. 20. Which may feem to take off the objected inconvenience. For if fo be that the Waters were to bring forth Fifhes, before they exifted, they could not lack agreeable Manfions upon their firft

emergency

emergency into being: inasmuch as the same Element was to afford them habitation, from whence they derived their production.

But grant that the Waters were to be productive of Fishes. Yet they might not be so *prima vice*, at the very first. Or if they did then help towards producing them; it could be only by yielding a rude kind of matter, out of which they might be formed: such as *Adam*'s Rib was for the making of *Eve*. And therefore though *GOD said*, at the Twentieth Verse, *Let the waters bring forth abundantly the moving creature that hath life*; Yet in the next Verse it is said, That *GOD CREATED every living creature that moveth, which the waters brought forth abundantly.* Where, if, ברא, *created*, does not denote GOD's making them out of pure nothing, (according to the rigid School-notion of *Creation*) yet it signifies (which is the *lowest* sense of the word) that he made them *ex materia prorsus inhabili*, out of matter of it self (till the Creator chang'd and dispos'd it) altogether unfit for such an use. So that albeit the Waters brought forth Fishes, yet they did not do it, by any *vis plastica*, *formative power* of their own solely; but so far as they afforded general (and naturally inept) Materials for their composition.

And, in some sense, the Waters (we know) have brought forth Fishes ever since. That is, by cherishing their *Spermata*, or *Spawn* committed to them. For they receiving those young and tender rudiments of life (upon their first ejection or exclusion) into their liquid Wombs; do nurse up the naked and imperfect Seminals, through the several Stages of an incompleat γένεσις, or *Birth*; till they arrive at animation and maturity. But then this implys that the First Fishes came into being by an extraordinary way; and could not be produced as they are now; because there were

non

none before them, none to propagate them, by casting forth such spermatic Principles. The aboriginal ones (as I may call them) for this very reason, must be made by GOD's immediate hand. Though whether he made them out of nothing, or out of watry Materials; is all one as to our purpose. For either way it was absolutely necessary, that Rivers should be extant as well as they; that so they might be in a readiness to receive them.

But now according to the Hypothesis under consideration, the Fishes of the two, must exist first, if the Creation (as I hope we have proved) were perfected in Six Days. For they were made upon the *Fifth Day*, says *Moses*, and how could there be Rivers so timely according to this new contrivance? The Sun it self was created but just the day before. And so what a work must here be done, to make Rivers coetaneous with the Fish we speak of? The beams of the Sun must have pierced into the Earth, and that so deep as to have reached the Abyss. And from thence plenty of Vapours must have been exhal'd into the Air. And these Vapours being upon the wing must have taken their flight as far as the Polar Regions. And there they must have been condensed into Rains. And these Rains must have made Bodies of Waters. And these Bodies of Waters must have been so great, as to have flowed along, through or against all obstacles. And these Floods must have been so violent, as to have hollow'd out Chanels for themselves all the way they went. And all this in one days space. Otherwise there must have been no Fishes made. Or they must not have been made, when GOD says they were. Or when they were made, there must have been no suitable Receptacles for them. For as for the Waters of the Abyss, they could by no means serve

Q for

for this use, as will appear in the Sequel of our Discourse.

4. But we are to pursue this matter farther yet. There could be no Rivers *in due time*; *that* has been evidenced. It is next to be proved that there could be *none at all before the Flood*.

How Rivers were first made, we have been instructed, by Rains descending from above. But whereabouts were these to fall? *In the Frigid Zones, or towards the Poles*, we are told; and * the Scheme representing them, shows as much. But then, methinks, they should have been in great danger, yea, under inavoidable necessity of Freezing. For the Sun (according to this Hypothesis) moving always in the Æquinoctial, before the Flood; he would constantly have been as remote from those raining Regions, as he is now from us in the depth of Winter, when he runs through *Capricorn*; or which is all one, when the Earth traverseth the opposite *Sign*. And there being *then*, no such Clouds as *now*; nor yet any Seas, by their foggy Vapours to mitigate the keenness of the Air; nor any Hills or Valleys, to cause a warmth by confused and irregular reflections of the Sun-beams: the Frosts within the Polar Circles, must needs have been exceeding sharp and terrible. And so the Fountains that should have fed the whole World with Water, would have been fast sealed up.

* Theor. pag. 227.

And then if the Earth were of an Oval Figure (as this Hypothesis affirms) grant but its Diameter to have been the same at its Æquinoctial parts *then*, that it is *now* (as in reason it must be greater, because it is fallen in since, and so grown less) and this would have set its Poles a great deal farther from

from the *Æquator*, and ſo from the Sun. For to inlarge a Circle, into an Oval Figure, its *area* muſt be made a quarter as big again at leaſt one way of its Diameter, as it was before.

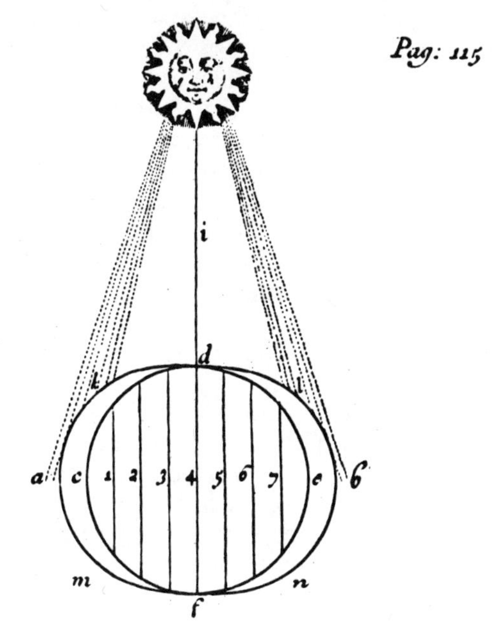

Thus, if the Circle *c d e f* be divided into Eight parts, by the parallel lines 1 2 3 4 5 6 7: we ſhall find that the two Arches *a* and *b*, forming the Circle into a *moderate* Oval; will at the points *a* and *b*, include ſuch ſpaces between themſelves and the ſides of the Circle *c*, and *e*, as ſhall be equal in breadth, to any two ſpaces betwixt the equidiſtant Parallels. According to which proportions allowing the Earth to be 7000. Miles in Diameter (though the true meaſure of it makes it more) and then adding a fourth part to it, to render it Oval, *viz.* 1750. Mile thickneſs: the Earth at each Pole, muſt bear above fourteen Degrees Latitude, or near nine hundred Miles extent, more than if it had been exactly round. And that *Hypotheſis* which removes its Poles ſo much farther from the Sun; muſt alſo allow the cold thereabouts to be proportionably augmented. And though in the hundred and fourth Degree of Latitude (as we muſt call it) on each ſide of the Æquator; that is, at the very Poles, there might have been perpetual day; the beams of the Sun reaching *a* and *b*, the two Poles of the ſuppoſed Oval Earth; and illightning them continually: yet his heat in thoſe places, muſt needs have been exceeding languid; foraſmuch as his Orb would always have been half above, and half under the *Horizon* to them. This will be clear from the Scheme, if we do but conceive the line *i d f* to be the Æquator, and the Sun ever moving directly in it. For then it muſt divide him into two Semidiameters *g* and *h*, at all times conſpicuous at the Poles reſpectively. That is to ſay, the Semidiameter *g*, at the Pole *a*; and the Semidiameter, *h*, at the Pole *b*. But then the Sun's being thus halved, muſt of neceſſity be a mighty diminution of his influence, eſpecially at ſo extraordinary a diſtance. It would have rendred his warmth more faint, than it is with us in the Winter Solſtice, when he is juſt a Setting, or half ſet.

But our buſineſs is rather to enquire what the temperature of the Air would have been, nearer to to the polar Circles; where these Rains are conceived to have fallen. Now if these Regions were as remote from the Sun, as we are when he is farthest from us; the Air must have been every whit as freezing there, as it is with us in the very dead of Winter. And that they were some degrees farther from the Sun, I think we need not question. For when the Sun is gone farthest from us, he reaches but to twenty three degrees and an half of Southern Latitude: which added to our fifty two of Northern, the whole amounts but to seventy five and an half. But granting the Earth to have been stretcht out to that length, to which its oval fashion would have extended it; and the supposed dripping Countries in the first World, might easily have been farther from the Sun (and consequently colder) by several Degrees.

In case it be opposed, That nights with us, when the Sun is retired to his utmost point in *Capricorn*, are some hours longer, than they could be in the prediluvial State; and that this might so far strengthen the Cold, as to make it superior to what it could be in the wet Regions we speak of: I answer, though our Nights be somewhat longer; yet we now dwell among Clouds and Seas, which do very much bemist and thicken the Air; and so make it warmer than it could be in the primitive World, where neither of them were to be found at the rate we have them. And truly the perpetual absence of them, must needs have made the Air more severely nipping in the Frigid Zones *then*, than it is *now*. Especially they being shot out so far from the Sun, by virtue of the oblong figure of the first Earth. For even as the Earth is now of a Globular make; the Rains might have fallen in the Frigid Zones for ten Degrees latitude, or six hundred Miles together, and yet (on (on the one ſide have been five Degrees diſtant from the Poles themſelves; and on the other ſide) have been ſeventy five Degrees diſtant from the Sun in the Æquinox; which is as far (to half a Degree) as he is ever remov'd from us. But then if we add better than fourteen Degrees more to each Pole, upon accompt of the Earth's Oviformity; the Rains muſt be removed a great way farther from the Sun ſtill (perhaps the whole fourteen Degrees) into Climates moſt horridly cold and freezing. And though there would have been conſtant Day about the very Poles; yet in this Oval Earth, there would have been as much Night in the preſumed rainy Regions; as in any other part of it whatever. For ſo we may obſerve, that thoſe rays of the Sun, which fell upon that Earth, ſuppoſe at *k* and *l* (whereabouts according to the Hydrographic * Scheme in the *Theory*, we may imagine the Rainy Regions were) could not illighten the oppoſite ſide of it at *m* and *n*, till ſuch time as thoſe points were turned to him, which they could not be ſooner than the point *f*; where it muſt have been of the biggeſt circumference, meaſuring it in way of Longitude.

* Page. 227.

Indeed it muſt be owned that it is not the Sun's diſtance in Winter, which does only or chiefly make our Climate ſo cold; but the oblique falling of his beams on the Earth. So that inſtead of his retreating Southward forty ſeven Degrees (the whole ſpace between the Tropics) were he at the time of his entring into *Cancer* (when he is neareſt to us) but elevated directly as many Degrees, or removed only perpendicularly from us: our Winter (if any) would be very moderate, becauſe his beams would be reflected in the ſame Angles as before. But his receſſion from us being in way of latitude or declination; his Rays muſt fall the more obliquely upon the Earth.

From

From which kind of incidence it comes to paſs, that they rebound in obtuſe Angles, and the heat which ſhould be cauſed by more direct reverberations, is impaired. As alſo many of his beams are reflected by the *Atmoſphære*, another way, and come not at us at all. But then the Sun being farther diſtant from the rainy Regions in the prædiluvian Earth; his beams muſt have fallen more obliquely upon them ſtill; and ſo the cold muſt have been greater there, becauſe his influence was leſs. And therefore what can be thought, but that the Dewy Rains (if any could have been in thoſe parts) ſhould either in falling have been turned into Hails; or if they fell in Water, have been frozen into Ice. And ſo inſtead of ſtreaming along and refreſhing the Earth, they muſt have ſtood congeled into Mountains. Eſpecially if we conſider that extremely cold hanging Miſts muſt have always incircled thoſe Regions above; and ſo have ſhut out that ſorry kind of influence, which might have been derived from the ſo remote and feeble Sun.

It may a little inforce what has been ſaid, that all who have held (with the *Theoriſt*) the Torrid Zone was uninhabitable by reaſon of heat; ever believed that the Frigid ones were ſo, through extremity of cold: as *Ariſtotle*, *Cicero*, *Strabo*, *Mela*, *Pliny*, and others.

To which add, That ſeveral Navigators, attempting to find out a nearer courſe to *China*, have been frozen to death. Yet they ſailed nothing ſo far Northward, as the rainy Regions in the Oval Earth, muſt have lain. Though without queſtion they choſe the moſt ſeaſonable time for the Enterprize; I mean when the Sun was on this ſide of the Æquator: where now he may advance (though he could not do

ſo

so (says the *Theory*) before the Flood) twenty three Degrees and an half; which on Earth we reckon about fourteen hundred Miles.

Nor is what *Mercator* remembers touching *Nova Zembla*, impertinent to the Case. *Here the Air is very sharp, and the Cold most vehement and intolerable.* And again; *their Tents are covered with Whales skins, the Cold being continually very sharp in these parts. Their drink* (the Geographer goes on) *is warm blood of wild Beasts, or else Ice water; there are no Rivers or Springs, because the violence of the Cold does so shut up the Earth, that Springs of waters cannot break forth.* And where Rivers cannot flow out of the Earth for Frost; surely they cannot fall down from Heaven. Yet this Island is extended but from the Seventieth to the Seventy sixth Degree of Northern Latitude, or thereabouts. *Speed* also informs us, that the Isles of *Shetland* in the *Deucalidonian* Sea, are *ever covered with Ice and Snow.* Yet *Ptolomy* placeth them but in the Sixty third Degree of Latitude; which is a good way on this side of the Arctic Circle. *Heylin* also says of *Island*, that *it is a damnable cold Country.* And *Blaeu* reports of the *Frigid Zones*, * *Perpetuum istic horridumque est frigus, There is perpetual and horrid Cold.* Lastly, the *Theorist* himself so far agrees with us, as to own that the *Frigid Zones* in the first Earth were uninhabitable; and that by reason of † *Cold*, as well as *Moisture*.

* Instit. Astron. *de usu Glob.* c. 4 § 2.

† Pag. 242.

CHAP.

CHAP. VI.

1. Another *Exception against the* Hypothesis; it would have drowned the world, though Man had not sinned: 2 Or though Mankind had been never so penitent. 3. *Which would have* reflected *upon* Providence: *and* imboldened *the* Atheist.

1. WE are taught from above, That * GOD *brought in the Flood upon the World of the UNGODLY.* That is, it was a Judicial act of His, and a just revenge which he took upon the impious. They had grievously offended and provoked His MAJESTY, by very great and epidemical Sins. For as we read in the Sixth of *Genesis*, *the wickedness of Man was great, and every imagination of the thoughts of his heart was only evil continually; and all flesh had corrupted his way* before him. Whereupon the HOLY GHOST speaking of God ἀνθρωποπαθῶς, after the manner of Men; declares that he was grieved at the heart to see this. And such was the grief he conceived, that He repented He made Man. And so vehemently did He repent of making him, that He resolved to destroy him again. And not only *him*, but most of his fellow creatures with him, made in good measure for his use and benefit. And not only *them* but the Earth it self in some sense, which had been the scene of his vanity and unrighteousness. And at length He decrees and proclaims aloud, that the Instrument of this fearful general destruction, should be a Deluge of water, *Gen. 6. 17.* So that nothing can be more clear, than that the Flood was a punishment of Man's sins, and was design'd and sent

* 2 Pet. 2. 5.

on purpose to be so. The Consequence from which is, that if Man had not transgressed; the Earth had never been so lamentably drowned.

But here then the truth of the *Hypothesis* we are upon will come into Question: in that it would have let in the Flood upon the World, though it had not been ungodly; though Men had been never so innocent or upright. For if the Earth had been formed, as is above supposed; it must have been of the same structure that is there phansied. It must have held the same situation to the Sun, and the same motion about it. And the Sun must have had the same power over the Earth, and the same effects upon it. It must have pierced it as deep, and parched it as much, and ripened it as fast for disruption as ever. The time of which being once come, down it must have plunged into the *Abyss* below; and all living upon it, must have sunk and drowned together with it self. No Natural Causes could have had the least regard to moral integrity: but on they would have driven in their appointed Courses, till they had come to the Tragical event we speak of. So that had all the Sons and Daughters of Men, been as pure and bright, as they could possibly have dropt out of the Mint of Creation; they must still have perisht without pity or remedy.

And so what would have become of the first Covenant with *Adam*, in case he had stood? For by such a Fatality as this in Nature (not unlike to absolute Decree in Divinity) his Posterity must have died, though he had not sinned, nor they neither. Which would have been a strange and unparallel'd severity, and such as did never issue from GOD. *Tophet* indeed is prepared of old; and there are endless and intolerable torments beyond this life. But none need suffer them unless they please. For still

we

we must be authors of our own misery if any betides us. And if our happiness chance to be blown up at last, the Train that does it, must be laid and fired by our selves. But by this *Hypothesis*, the Race of Mankind must have been wofully undone, though they never deserv'd it. For the primitive Earth had that in it, which we have; Frailty in its very Nature or Constitution: And in the ordinary setled Course of things, must necessarily have been dissolv'd and delug'd.

2. And if purest Innocence must have fared thus ill; *Repentance* for certain should have sped no better. *That* I add for this reason. *Noah*, we know, was Δικαιοσύνης Κήρυξ, a Preacher of righteousness. And that not only in a vocal way; but by his religious and exemplary life. Yea, more than so, his building the Ark was a Mechanical Sermon to the World; and perhaps of an hundred and twenty years long. *For* in the same * Chapter where GOD denounc'd the Sentence of Inundation, and commanded *Noah* to prepare the Ark; He determin'd and declar'd, that *the days of Man* (that is, before the Flood was to come) *shall be an hundred and twenty years.* And such a way of Preaching, and of such a continuance; in reason should have wrought with that stubborn Age, beyond the most elaborate and pathetic Discourses. And GOD seems to have expected no less. For because it did not, His Holy SPIRIT has clapt a black Brand upon them, and markt them out for incorrigible and ungracious Wretches. *Who were disobedient in the Days of Noah*, κατασκευαζομένης κιβωτοῦ, **while the Ark was a preparing.** Disobedience in that juncture, under such a warning against it, and motive to the contrary; was such a disobedience, as

* Gen. 6.

as for circumstances of aggravation could hardly be parallell'd in that World. It deserv'd to be recorded as a standing testimony against them that were guilty of it, and as an eternal monument of their base unworthiness. For it was no other than the fruit of contumacious refractoriness, and bespeaks them arrived at the height of obstinacy, and a most consummate vitiousness.

But put case the Sermons of this mighty Preacher, had wrought so kindly and effectually, as to have turned Mens disobedience into true repentance: would this have altered the State of Nature, or put a stop to its fatal tendencies? Not in the least measure. Still the World would have stood in its Original frame; and a change in the minds and manners of people, would have made none in the Physical Course of things. So that unless God had interpos'd and by His immediate hand, given a timely check to Natures Wheels; they would have run directly into this watry ruine, and what should have kept the sincerest penitents out of it? For to suppose that GOD ingaged so far, as to support the Earth by strength of *miracle*, to secure it from the Flood; would be as great a flaw in the Philosophy of this new *Hypothesis*; * as it is thought to be in the Divinity of the Old one, to hold the Deluge was caused by Creation of Waters, and then dried up by annihilation of the same.

* *Read* 18th, 19th, and 20th, *pages of the* Theory.

3. And yet if Omnipotence had not *miraculously* upheld the Earth,) supposing its Inhabitants righteous or penitent) it would have fallen heavy upon GOD Himself. So heavy, as to have crusht the Reputation of his Providence extremely. For it would have recoiled so rudely and violently upon its Goodness; and

and so shamefully Eclipsed and blasted its Justice: as to have brought its very *Being* into question. And we may certainly conclude, that an *Hypothesis* of this Nature, which would weigh out the Portion of Men, with so inequal a Ballance, as to make ruine the lot of a righteous or repenting World; instead of gagging or silencing the pragmatical *Atheist*, by a more clever Explication of the Deluge: would open his Mouth wider, and but oil his virulent and sawcy tongue, to run more glibly and rantingly on, in his tremendous way of extravagance. For what can more encourage so wicked a person, than to disparage and lessen GOD's Goodness and Equity? And how can those Attributes be more disgraced and diminished (in the Judgment of an *Atheist*) than by supposing that in the Works of His Providence (through the whole *Series* of which he could look with a clear and easy prospect; and so nothing of oversight could mingle with them) He laid a cruel Train of inavoidable Death, for Millions of his Innocent or Penitent Creatures.

How little this would have comported with those His illustrious and Cardinal Properties, and how much it would have blemisht and dishonoured them; we may guess from hence: in that when he was minded to overthrow *Sodom*, and in his Holy Agents was come down from Heaven on purpose to do it; He would have spar'd it for the sake of Ten righteous Persons. And truly if He had destroyed the righteous with the wicked; He must have done a thing (in the sense of *Abraham*) not at all agreeable, to the Integrity of the Judge of all the Earth. (Especially in those Ages, when Spiritual Encouragements to GOD's Service, and recompences of it, being not so frequently dispensed; and the Eternal ones not

so

so fully revealed: the Divine favour was more commonly measured and expressed to Men, by temporal and outward Blessings and deliverances.) And therefore that He abhorred such inequitable Dealings, he was pleased to evidence by the contrary Procedure. For when He consumed that accursed Town, he saved just *Lot* by the Ministery of Angels. Nor could He endure that *Noah* should perish, being righteous; but took particular care for his wonderfull preservation, when the whole World, besides Him, and his Family, was drowned.

But then so much less reason there is, to admit *this Hypothesis*; for that it makes the Earth (at first) of such a Form; and puts Nature into such a Frame; as would have involv'd Mankind in most horrid Destruction. And not only so, but moreover makes Providence accessary to their Perdition; yea, the principal and sole Contriver of it, by making the place of their Habitation, a perfect Trap to vast multitudes of them: whereby (without a Miracle) they must certainly have been taken and quite undone, had they been never so pure, or never so penitent.

Should it be suggested, that GOD foresaw the impiety and incorrigibleness of Men; and so in way of just judgment, ordered Nature, and timed the Earth's Dissolution accordingly: this would give little satisfaction to the *Atheist* (the *silencing* of whose *Cavils* the *Theory* seems to aim at.) For he would take it at best, but for a smooth Evasion, or a slim Subterfuge; or for a sorry kind of Fetch to help the *Hypothesis* at a dead lift. Nor need we doubt but a *Lucian* or an *Hobbs*, would raise as considerable Objections against this *New way* of explaining the Flood, as against the Old one. And would insist as tenaciously upon that *Particular* now mentioned, and cavil

cavil as much, and as juſtly at it; as at the difficulty or unſolvableneſs of any ſingle *Phænomenon*, in the way of its *uſual* Explication.

CHAP. VII.

1. *Saint* Peter's *words alledged in* favour *of the* Hypotheſis; inapplicable *to that Purpoſe.* 2. *Wherein the* ſtreſs *of them ſeems to lie.* 3. Seven *other* Allegations *out of Scripture, of* no Force; 4. *As being* Figurative, *and ſo not* Argumentative. 5. *Which* Tycho Brache *not minding, it gave occaſion to his* Syſteme.

1. TO countenance the Formation and Structure of the Earth aforeſaid, the Ingenious *Theoriſt* has call d in ſeveral Divine Authorities. And it being attempted to authenticate the *Hypotheſis*, by Allegations of that nature; it is but neceſſary that we take notice of them, and ſhow their invalidity.

The firſt is cited out of the Second Epiſtle of *S. Peter*, and runs thus. * *For this they are willingly ignorant of, that by the Word of GOD, the Heavens were of old, and the Earth ſtanding out of the water, and in the water; whereby the World that then was, being overflowed with water, periſht. But the Heavens and the Earth that are now, by the ſame word are kept in ſtore, reſerv'd unto fire againſt the day of judgment.* Where, it is thought, † *the Apoſtle doth plainly intimate ſome difference, that was between the Old World, and our preſent World, in their form or conſtitution; by reaſon of which difference that was ſubject to periſh by a deluge, as this is ſubject to periſh by conflagration.* To wind his words into a favourable compliance with this ſenſe, ſome ſpecious

* Chap. 3. verſes 5, 6, 7.

† Theor. pag. 45.

ſpecious * offers are made. But inſtead of applying anſwers to each of them in Particular; we may ſhorten our work by obviating them with *one* general Obſervation touching the Paragraph, which is this: There is a Clauſe in it, that will by no means ſuffer it to be interpreted the *Theoriſt*'s way. Namely, *this they are willingly ignorant of.* And of what were they thus ignorant? Why, of the Nature of the firſt Heavens and Earth, and of the alterations that befel them at the time of the Flood. So we are aſſured. † *The Apoſtle tells them, that they are willingly ignorant of the firſt conſtitution of the Heavens and the Earth, and of that change and diſſolution which happen'd to them in the Deluge.* But if St. *Peter* meant theſe things, I dare boldly ſay, that his charge was too ſmart and heavy upon the Men; yea, falſe and unreaſonable. For though *ignorant* of the things they might well be; yet how could they be WILLINGLY ignorant of them? Muſt not that be hard to make out?

* Read the L. Biſhop of *Hereford's Animad. Sect.* 1. a'moſt throughout.

† Theor. p. 47, 48.

Let us try, but as to one of the mentioned heads, the FORM of the Earth. By what means ſhould they have come to the knowledge of *that*, though they would never ſo fain have done it? GOD had not reveled it, nor had Man apprehended it. And how then could their ignorance in the caſe be *wilfull*? In what Books was this Form of the Earth recorded? Or what lively Tokens or Monuments were there of it? Whence ſhould they have gathered it? Or where ſhould they have met with Intelligence concerning it? To ſay that Hills and Valleys, and Mountains and Rocks; that the Clifts of the Sea, and its Deeps and Chanels; that the rugged and broken Surface of the Ground, or any thing of that nature, might have informed them of it: would be but

but wild and extravagant talk. For beſides that theſe Scoffers whom the Apoſtle reproves, had no reaſon to believe, that the aforeſaid *Phænomenaes* were marks or Footſteps of a ruinated Earth: ſo if by chance they had phanſied them ſuch, they might ſtill have been far from a right *Idæa* of its ſuppoſed primitive frame. A man may view and review an heap of Rubbiſh, which was once an houſe, very long and often; and yet be never the more able at laſt, to pronounce what Model the Fabric was of. In like manner, the moſt curious Surveys and reiterated obſervations of things, in that confuſed poſture wherein the Earth preſents them to the eye; could never have led thoſe the Apoſtle diſputes againſt, into a right apprehenſion of this its Figure, which the *Theory* makes it of before the Flood.

Had there been fair Indications of ſuch a Form, why did they not direct Men into an *earlier* Diſcovery thereof? For touching it we find not one word in Antiquity. Yet Mountains, and Rocks, and the like *Deformities* in Nature (as we are taught to think them) were altogether as viſible ever ſince the Deluge, as they are now. And when none of the moſt ſearching prying minds; none of the moſt buſy, intelligent Speculators; were ever ſo quick-ſighted as to decry this Form of the Earth; from the aforeſaid (imagined) Irregularities, or any other hints or Characters of it: it was certainly a thing too obſcure, to fall under the notice of thoſe Herctical Mockers, deſervedly reprehended by the HOLY GHOST. But then how could He rebuke them, for being *wilfully* ignorant of it, it being ſo very dark a Myſtery?

Even by the *Theoriſt*'s own confeſſion, this Doctrine was always abſtruſe, and ſuch as the Wiſeſt

S * Philoſophers

* Philosophers did never hit upon. They never knew of a Paradisiacal Earth themselves, nor did they ever speak any thing of it to others. And when it was thus secret, and hidden from all learned Men; why should the HOLY SPIRIT (I say) tax these Scorners, with *wilfull ignorance*, for not understanding it? Who, however they might abound with *conceited* knowledge (as the name, *Gnostics*, which they arrogated to themselves, imports) were but pitiful Sciolists. † The *Theory* also affirms, that *Paradise* and the *Universal Flood*, were by *length of time*, *and the changed face of nature*, *so much obscured*; that if *holy Story had not minded us of them*, we should not only not have known them, but *never have thought of them*. And if the Flood had been utterly buried out of mind, and might never have come into the *thoughts* of Men, if Scripture had not kept it in memory: then what hope of understanding, that it was occasioned by such a form or Fabric of the Earth, as the *Theory* has invented; unless the same Scripture minds us of that also? But because it does not, how could the Persons whom *S. Peter* reproves, be *wilfully* ignorant of the *Phænomenon*?

* *Nullus enim Philosophorum, sive veterum, sive recentiorum, cujuscunque Sectæ, unquam animadvertit, aut ex causarum contemplatione invenit primam telluris faciem fuisse Paradisiacam.* Theor. p. 2.

† —— *Quibus temporis longinquitas, & mutata Naturæ facies; tantum obscuritatis attulisset, ut nisi excitati ab historia, sacra de iis forsan nunquam cogitassemus.* Pag. 116.

Wilfull Ignorance, is that which GOD blames, and which is really faulty upon *our* account: which we carelesly rest in when we might come out of. When Men might have means of knowledge, but will not seek them; or when they actually have them, but will not use them; but in the midst of proper helps to science, sit down and chuse to acquiesce in Ignorance; this is *wilfull* and affected. But these were not the circumstances of those, whom we find to have been Objects of the Apostolical Censure. They were so far from standing fair for acquaintance with this structure of the Earth, or from being in a probable

bable way to the knowledge of it; that they were next door to an utter impoſſibility of ever attaining it, ſuppoſing it had been real. For their Minds were ſet (I may ſay) with a contrary Biaſs, and it was morally neceſſary that they ſhould be drawn the other way. For the whole World was of that Judgment, it is of *now*, and which theſe Mockers were of *then*: and why ſhould they differ from all people then alive, or that ever lived? *It hath been generally thought or preſum'd* (ſays the * *Theory*) *that the World before the Flood, was of the ſame Form or Conſtitution with the preſent World.* And how could they help ſwimming with the general Stream? Yea, which is more, the Opinion was as *Strong*, as it was *general*; and ſtood very firmly in Mens apprehenſions, they thinking it built upon Scripture Grounds. For that ſpeaks of Seas created in the *beginning*, and of Mountains covered with Water in the Deluge. And all agreeing, that the Seas mentioned by *Moſes*, were no other than thoſe which are now extant; and that the Mountains ſo covered, were præexiſtent to the Flood: the preſent face of things, which is † preſumed of good uſe to evince the Earth was of another Form once; became a great Argument to perſwade theſe Scorners, that it was always of the Form which it now bears; and a means to fix them in that Perſwaſion. And when their condition was ſuch, as to be deſtitute of the knowledge of the Form of the Earth; and the moſt likely means they had to help them to it, would rather have run them upon the contrary belief, and rivetted them faſt in it; there could be no reaſon why they ſhould be charged with *Wilfull* ignorance of the thing. And if they could not upon juſt grounds be charged with *Wilfull* ignorance of the Form of the Earth; then neither with the like ignorance

* Page 276.

† *It will be found, it may be, upon a ſtricter Enquiry, that in the preſent form and conſtitution of the Earth, there are certain marks or indications of its firſt State.* Theor. p. 8.

of the Constitution of the Heavens, and of the Change and Dissolution that happened to either: they being things as much in the dark, and as far removed out of the way of their notice. Let us but just point at each of them.

The whole Superficies of the Terrestrial Globe was entire and continued, smooth and even, regular and level. No Lake nor Sea, no Rock nor Island, no Hill nor Dale, was any where upon it. But as the Earth was made of two distinct Orbs; so betwixt its outward Orb of an Oval figure, *and* that within; was the great Body of the Waters lodg'd and shut up so close, as to hold no commerce with the open Air. Such in gross was the Form and constitution of the first Earth.

The Sun piercing through the outward Orb of the Earth, drew up (chiefly about the Middle parts of it) great quantities of Vapours, out of the Abyss. Which Vapours directing their Courses in the Air, from the Æquinoctial to the Polar Regions; they were there condensed into Rains, to furnish the World with Rivers. But these streams of Exhalations flowing continually through the Aereal Regions; made them exceeding watry. And such in general was the Form or Constitution of the Heavens.

The Sun moving always in the Æquinoctial, the Earth grew extremely dry about the Æquator, and full of Chaps; which rendred it more weak and brittle in its exterior Orb. Which Orb being fill'd with Vapours within, raised by the penetrating heat of the Sun, was still more apt to be blown up and broken. At length being able to hold no longer, it flew in pieces, and down it fell into the Deep beneath, sinking till it rested on the Orb below. Such in short, was the Earth's Dissolution.

By

By the fall of *that* into the Waters under it, *they* were forced violently to fly up aloft; and ſurging and raging in a tumultuous manner, the great and fatal Deluge was cauſed. Hence alſo Seas and Lakes aroſe, while the watry Element abating of its fury, quietly retired into ſuch hollowneſſes as were ready to receive it. And whereas the external Orb of Earth, was ſo much bigger than that within, as to contain the whole Maſs of Water in its Cavity; and ſo could not poſſibly ſurround and ſit cloſe to the inward leſſer one, in an orbicular faſhion about it; but ſeveral of its parts in ſeveral places, were fain to ſtand erect inclining, *&c.* theſe various Prominencies of different ſizes, ſhapes, and ſituations; made Mountains and Rocks of all ſorts. But the Outward Earth being thus diſſolved, and fallen as low into the Waters as it could; it was no more liable to a general Flood, but was certainly put paſt that danger for ever. And thus *Its* Form and Conſtitution was altered.

Now the Sun alſo running a new courſe about the Earth, by reaſon ſhe had changed her old Poſition; and the Abyſs being diſordered by the Diſruption of the Earth, and its falling into it: Vapours could no longer be drawn out from thence as they uſed to be, nor fill the Aereal channels with ſtore of Exhalations. And ſo they growing dry, the watry Complexion of the Heavens periſh'd; and *Their* Conſtitution was changed alſo.

Such in brief (ſo far as we are concern'd to note at preſent) was the Form and Conſtitution of the Heavens and the Earth; and ſuch the changes they both underwent, as the *Theory* teaches. If therefore the Parties *S. Peter* reproves, were blamed for not knowing *the firſt conſtitution of the Heavens and the Earth*,

Earth, and that change and dissolution which happened to them in the Deluge; their ignorance of *those Particulars rehearsed* must be the Summ of their Charge. But then all those things being perfectly new; such as neither *Pythagoras*, nor *Plato*, nor *Aristotle*, nor *Zeno*, nor any Philosophers of any Sect or Age, did understand and declare: how can it be thought that silly *Gnostics* or Pseudo-Christians could be acquainted with them? And yet if they could not, then neither could they be condemned of *wilfull* ignorance of them; nor can the Text be applied to the *Theory*'s *Hypothesis*.

2. But if this were not S. *Peter*'s Drift; if it were not his intent to rebuke them for their ignorance of these things; what then could be the scope of his correption? I answer. Though he could not give them this gird for their being ignorant of the *Flood*; yet he might do it properly for their being ignorant of the *Chief Cause* of it. Ignorant of the Flood they could not be, it was a thing so well known, and so generally received in the Church. That the Heavens were of old, and the Earth standing out of the water and in the water; and that in their standing thus, the then World was overflowed with water, and perished: they could not be chargeable with ignorance of this. But the stress or Emphasis of the Apostles charge lies here, that they were ignorant of its being done *by the Word of GOD. The Heavens were of old*, says he, *and the Earth standing out of the Water*, and in the water, τῷ τοῦ Θεοῦ λόγῳ, *by the word of GOD*. And then it follows, δι' ὧν, *by which* (that is, by which situation, and by which word of GOD causing it) *the World that then was being overflowed with water, perished.* So

So that *this seems* to have been their fault, that they had not a true notion of the principal Cause of the Deluge. But through their own heedlesness, were much in the dark and mightily to seek as to that particular. They did not know, because they would not; that GOD brought it in by the word of his power; or in pursuance of that righteous decretory Sentence, denounced by him, *Gen.* 6. They were of opinion (as others have been) that the Flood was a meer *casual* thing; and that the hand of GOD was no otherwise in it, than in the purest contingent Calamities. Or else that it proceeded wholly from Nature and Second Causes; as from the Conjunction and influence of watry Planets. However they might think it was of larger extent, and longer Duration; they might ascribe it to no higher Cause, than some do the Flood of *Ogyges*, that happened in *Arcadia*; or that of *Deucalion*, which drowned *Thessaly*. Concerning the latter of which, *Lucan* thus phansied.

> *Deucalioneos* fudisset *Aquarius* imbres.

> Aquarius *'twas that made those rains pour down,*
> *Which in* Deucalion's *time the Earth did drown.*

He plainly imputed it to Astral efficiency, or the force of the heavenly Constellations. Now if these Men thought thus vainly of the general Innundation; and knew it not to be the Effect of the special Providence of GOD: they were grosly ignorant in the Case, and this their ignorance was grievous *wilfull*, and deserved the holy reproof they met with. For had they but consulted the Story of it, and considered what *Moses* says concerning it; they would soon have perceived, it was the direful issue of of divine power and justice, and came not by the influence of the Stars, but by the appointment of the DEITY.

And that to condemn this very ignorance, was the real meaning of S. *Peter* seems to be clear from its agreeableness to his aim or intention. Which was to prove the World's Conflagration upon perverse Men who question'd the same, and disputed against it by this Argument, that *all things continued as they were from the beginning* (whereby they hardned themselves against the Doctrine of the Conflagration, in which the Apostle threatned them with a dismal *Catastrophe.*) Now how does the Apostle answer and take off this? Why, by fetching a compass about in his Discourse, and by telling them (though not in these words,) to this purpose: that when the World was to be *drowned*, all things continued *then* as they were from the beginning; and Nature did not signify it beforehand, by any sensible observable Changes, because the work was not to be naturally done, but *by the Word of GOD* commanding and causing it: which to be ignorant of, was their great fault. And therefore that in their time, *all things continued as they were from the beginning*, ought to be no reason to them, that the World shall not be *Burned*; because it is not to be expected, that Nature should foreshow it by any prævious alterations: inasmuch as this *Burning*, is no more to be effected in a natural way, than the Deluge was; but τῷ ἀυτῷ λόγῳ, *by the same Word* which drowned the World, and by which *the Heavens and the Earth which are now, are reserved unto fire.*

Were it necessary in the least, after what has been said; it might here be noted, that the words are very capable of, and might properly be expounded

to

to *another* ſenſe. *This they are willingly ignorant of:* That is, *they are willingly* mindleſs or forgetfull of it. For ſo λανθάνει, may ſignify (which in the caſe the Apoſtle ſpeaks to, muſt be an hainous fault, and worthy of reprehenſion) and therefore a thing *forgotten*, is ſaid παραδιδόναι λήθη: and *Iſocrates* commending the actions of *Hercules* and *Theſeus*; ſays they were ſuch ὥστε μηδὲ τὸν ἅπαντα χρόνον δύνασθαι λήθην ἐμποιῆσαι, &c. *as no time could bring into oblivion*, or blot out of remembrance. Tho' ſtill there would be as little reaſon to charge theſe Heretics, with *wilfull forgetfulneſs* of *thoſe things* that the *Theory* would make the Text point at; as there is to check them for *wilfull ignorance* of the ſame.

3. Beſides this of *S. Peter*, * *other places of Scripture ſeem manifeſtly to deſcribe this ſame* (new) *form of the Abyſs with the Earth above it*; as we are told. But as all thoſe places may as *well* or *better* be applied to the Earth in its *preſent* form; ſo they can hardly be interpreted in favour of *this other*, without ſome kind of violence or abſurdity. *Theor.p. 86.

The Firſt occurrs *Pſal.* 24. 2. *He hath founded it* (the Earth) *upon the Seas, and eſtabliſhed it upon the Floods.* Where, על rendred, *upon*, does as properly ſignify, *by*. And ſo, He founded it *by* the Seas, and eſtabliſhed it *by* the Floods. Which *David* might the rather note, becauſe ſo much of *Paleſtine*, (where he lived) lay along by the *Mediterranean*. Though when our Learned Tranſlators turned the word *upon*; they made it ſpeak moſt true *Engliſh*. For where land lies *by* the Sea, we commonly ſay, it lies *upon* it.

But then on the other ſide, the Earth, according to its firſt (imagined) form, could in ſtrictneſs be founded neither *upon* the Seas, nor yet *by* them;

T because,

because no Seas were *then* in being, but only an *Abyss*. Should it be answered, that the Abyss is here called Seas, by a *Prolepsis*; I rejoin: Those Seas must then be called, Floods, by another *Prolepsis*. And so the advantage will be cast on *our* side. For in respect of the present form of the Earth, the words may be expounded most naturally, without a Figure: but in reference to the other form, they must not only be strained up to a Figure, but that Figure must be twice made use of. And which is very considerable, נהרות rendred, *Floods*; does signify *Rivers:* and so the LXX, and *Vulgar* do both render it. And though that sense falls in most properly with the present form of the Earth, as it is every where extended by Rivers; yet it can by no means hold with its *first form*, supposing it established upon the Abyss: for in that (allowing there were *Floods*) there could be no *Rivers*.

As to the next place, *Psal.* 136. 6. *Who stretched out the Earth above the Waters*, We need say no more than has been said already. It may as well be read *juxta aquas*, *by* the waters, as, *super aquas*, *above* the waters.

The Third place is, *Psal.* 33. 7. *He gathered the waters of the Sea as in a Bagg, He layeth up the Abysses in Storehouses*. Which, says the Theory, * *answers very fitly and naturally to the place and disposition of the Abyss which it had before the Deluge, inclosed within the Vault of the Earth, as in a Bagg or in a Storehouse*. But I say it sutes the present form of the Earth as well as it does the first: only this difference. The *Bagg* and *Store-houses*, supposed to be in the first Earth, were *shut*; but in this, they are *open*. Yea, it sutes it much better upon two accounts. For in the Earth as it is now, there are, אוצרות, *many Treasuries* or *Storehouses* of Waters (ac-

* Page. 86.

according to the Text) which has the word in the *Plural* Number. Whereas in the first Earth, there could be but *one*, before the disruption. And then the word כנר, rendred, *as in a* 𝔅𝔞𝔤, should be rendred, *as on an* 𝔥𝔢𝔞𝔭, as it is in the *English*. The *Theory* indeed faults that Reddition, as not making (a) *a true sense*. But in all likelihood, our Translators were in the right; for, נד, properly is an *heap*. And though, נאד signifies a *Bag*, yet (as (b) *Buxtorf* notes) where it is written without *Aleph*, it is not found in that signification; but signifies an *heap*. And so says (c) *Fagius*; and the same says (d) *Masius*. And therefore (e) *Schindler* renders, כנר, in this very place, *tanquam cumulum, as an* 𝔥𝔢𝔞𝔭. And so does (f) *Bithnor*, adding, That whereas the *Targum*, and LXX. render it, *a Bag*, it was because they read it, נאד. But נד (coming of נוד, *motion*; and so being *quasi rei motæ in unum congregatio, the gathering of a thing moved into one*) he will have to signifie an *heap*.

(a) *Ibid.*

(b) *Sine* א *pro utre non reperitur.* Vind. ver. Heb.

(c) *Uter est* נֹאד, *non* נד. in *Exod.* 15. 8. Translat. V. T. Collat.

(d) *Verùm* נֹאד *utrem significat,* נד *acervum.* in *Josh.* 3. 16.

(e) Lexic. Pent. *in Vocab.* נוד.

(f) Lyr. Prophet. *in loc.*

And whereas the *Theory* alleges, That the *Vulgate, Septuagint*, &c. render the word (g) in a 𝔅𝔞𝔤, *or by Terms equivalent*; yet granting that to be the only proper Reddition, it would make nothing at all to the *Theorist*'s purpose; another place of Scripture plainly defeats it. For *Psal.* 78. 13. we read in the *Septuagint*, ἔστησεν ὕδατα ὡσεὶ ἀσκόν: And in the *Vulgar, statuit aquas quasi in utre. He set the Waters as in a* 𝔅𝔞𝔤. Which not only makes the forecited Clause of the 33. *Psalm*, to be no manner of evidence of the Seas being inclosed at first; but moreover makes it a *Proof* of the clean contrary. For it speaks of the *Red-Sea*, and

(g) Page 86, 87.

and says it was *in a Bag*, as much as the fore-quoted Text can possibly be made to say, that the *Abyss* or Proleptic Sea was so: and yet at the same time it was not only *open*, (as other Seas are now) but much more open than ever. For it speaks of it at that very time when *Israel* passed through it, as the same Verse testifies. And whereas the *Theorist* notes, that *the Oriental Versions and Paraphrase*, render the word (as he does) *in a Bag*: I may affirm that the *Targum, Syriac, Arabic*, &c. render it so in the place I have alleged. But how little their Authorities will countenance his Exposition of the *Psalmist*'s words which *he* cites; and how little that Exposition will help his *Hypothesis* of the form of the Earth; may appear from the *Psalmist*'s words that I have cited. Which if they had been considered, might have damped *that thought*, which concludes the Paragraph belonging to that place of Scripture we have now spoken to. The thought is thus expressed by the *Theory*, * *I think it cannot but be acknowledged, that those Passages which we have instanced in, are more fairly and aptly understood of the ancient form of the Sea, or the Abyss, as it was enclosed within the Earth; than of the present form of it in an open Chanel.* But then that Passage in *Psal.* 78. 13. (being parallel to *Psal.* 33. 7. so far as we are concerned in it) must be acknowledged to be most fairly and aptly understood, of the *Red Seas being enclosed within the Earth*, when *Moses* and the *Hebrews* marched through it: and could that be?

* Page 87.

The next place is *Job* 26. 7. *He stretcheth out the North over the empty places, and hangeth the Earth upon nothing.* The same is as true of the *South* also; but the good Man living in this Hemisphere, the *North* was the nearer and more obvious of the two. And what could be more agreeable to the present

present Earth? For it having no visible sensible thing under it or about it, to shoar it up or support it; it may very well seem in common apprehension; and be said in the vulgar way of speaking, to be stretcht out upon emptiness, and hanged upon nothing. And so the Sun stood still upon *Gibeon*, and the Moon in the Valley of *Ajalon*: though the places were without the *Tropic*. And however *Job* in this Expression, might accommodate himself to the ordinary Phancy and speech of Men, while he represents the Earth as extended and pendent over an immense vacuity; yet (to cry quit with the *Theory*, which makes an illiterate Apostle, a profound Philosopher; let me say) in the truth of the Notion he was a perfect *Platonist*. For in this matter (whensoever he lived) he fully agrees with *Plato's* Doctrine. For *he* also conceived the *Earth to be hanged upon nothing*, as having no other Prop to sustain it, but its own figure and equiponderancy; by which it swims evenly in the Element about it. In testimony of this (and so of the mutual concent betwixt *Job* and him) let this Passage out of his *Phædo* speak. *(a) I am perswaded that if the Earth be but in the midst of the Heavens, it needs not the air, nor any other help of the like nature, to keep it from falling. But that a general equality of the Heaven in it self, and an even-poizedness of the Earth, is sufficient. For an Equilibrious thing placed in a sutable* (or similar) medium, *will not sway any way little or much; but keeping it self evenly* (balanced) *is free from inclination.*

Josh. 10. 12.

(a) Πέπεισμαι τοίνυν εἰ ἔστιν ἐν μέσῳ τῷ οὐρανῷ περιφερὴς οὖσα, μηδὲν αὐτῇ δεῖν μήτε ἀέρος πρὸς τὸ μὴ πεσεῖν, μήτ᾽ ἄλλης ἀνάγκης μηδεμιᾶς τοιαύτης. Ἀλλὰ ἱκανὴν γὰρ αὐτὴν ἴσχειν τὴν ὁμοιότητα τοῦ οὐρανοῦ αὐτοῦ ἑαυτῷ πάντῃ, καὶ τῆς γῆς τὴν ἰσορροπίαν. Ἰσόρροπον γὰρ πρᾶγμα, ὁμοίου τινὸς ἐν μέσῳ τεθὲν, οὐχ ἕξει μᾶλλον οὐδ᾽ ἧττον οὐδαμόσε κλιθῆναι. Ὁμοίως δ᾽ ἔχον, ἀκλινὲς μενεῖ.

But

But in this New *Hypothesis*, *Job*'s notion can have no place. For to say, the Earth, according to that, was stretcht out upon emptiness, and hanged upon nothing; would be notoriously false. For the *Theory* teaches that it * *rise upon the face of the Chaos*; † and *could not have been formed unless by a concretion upon the face of the Waters*: and that it had *the * mass of Waters* as a *basis or foundation to rest upon*. And so the Antediluvian Earth was no more stretcht out upon emptiness, and hang'd upon nothing; than an Arch is, when it is built upon its Center. And it was but just now that the *Theory* contended from that Passage in the 24. *Psalm*, that it was founded upon the Seas, and established upon the Floods. But how then could it be stretched out upon emptiness and hanged upon nothing? Or how can the two Texts, in the *Theory*'s sense, be reconciled?

* Page 58.
† Page 61.
* Ibid.

In case it be answered, That though the Earth at the very first was not stretcht out upon emptiness, and hanged upon nothing; yet in process of time it was so, when the Abyss was sunk in some measure, by reason of the huge quantity of Waters the Sun had drawn out of it; and so the Earth sat hollow about it: I reply in short, *Job* for certain meant no other than this present Earth: For in the very next Verse, he speaks of *thick Clouds* in which Waters were bound up, and they not rent. And such Clouds (according to the *Theory*) there could never be, till the first Earth was dissolved.

A Fifth place is *Job* 38. 4, 5, 6. *Where wast thou when I laid the foundations of the Earth? declare, if thou hast understanding. Who hath laid the measures thereof, if thou knowest? or who hath stretched the line upon it? Whereupon are the foundations thereof fastened?*

fastened? Or who laid the corner stone thereof? This likewise answers as properly, and perhaps more fully, to the present real form of the Earth; than to the other fictitious one. For GOD is here said to have laid the foundations of it. Which surely he may as properly be supposed to have done, in case he produced it by immediate Creation; as if there had been only matter and motion, and the power of gravity and levity in the Architecture of it; and so its formation had been meerly mechanical, as the *Theory* makes it.

And then *the measures* and *the line* that are here mentioned, do only imply that the Earth was made with fitting accuracy; of necessary and convenient, of regular and comely dimensions and proportions. And may not this Earth, in those regards be allowed to vie with that suppostitious one under debate? Yea, does it not in some things excel it? For though it has not the very same Elegancies, which that Earth had; yet it has other Imbellishments equal to them, if not beyond them. Indeed it has not that smoothness and entireness, which is pretended to have been in the first Earth. But then (which is more considerable) it has the raised work, of Hills; the Embossings, of Mountains; the Enamellings, of lesser Seas; the Open-work, of vast Oceans; and the Fret-work, of Rocks. To say nothing of those stately Curtains over-head (wanting heretofore) which are frequently drawn and flung open upon occasion; and sometimes curiously wrought and most richly gilt, even to admiration; far surpassing the goodliest Landskips, that ever were or can be painted: I mean the Clouds, which though they be things distinct from the Earth, yet having their beginning from the Earth; and from *this* Earth too

too (according to the *Theory*) in opposition to the other; are no improper instance of its out-vying it.

But, not to run out into endless Particulars, this Earth may compare with, and be thought to out-go *that imaginary* one, in Two general and chief things, *Comelin ss*, and *Usefulness*.

First, in *Comeliness*; For irregularities many times make a sort of Ornaments; and those ruggednesses and inequalities that are void of all exactness and order, do often pass for Beauties or a kind of Prettiness. But then more especially may they do so in the *Earth*, whose natural pulchritude is made up of such things as Art would call rudenesses; and consists in asymetries and a wild variety. And yet for an Earth, it is most beautiful and comely still. Thus an Urchin may be handsom in his kind, though he has not the beauty of a Dog; and a Dog, though he has not the beauty of an Horse; and an Horse, though he has not the beauty of a Man. And so is this Earth, though it has not the beauty of finer things in it, but only that which is peculiar to it self. For as the beauty of the Sun, lies in brightness and glory; and the beauty of the Sky, in clearness and serenity; so the beauty of the Earth, which is a different thing, does and must needs lie in very different instances, namely, in Seas and Lakes, and Islands and Continents; in Flats and Prominencies, and Plains and Protuberancies, and Hollownesses and Convexities; in smooth and spacious Levels in some places, and Hills and Mountainous Roughnesses in others. Whose careless diversifications, and interchangeable mixtures, as they mutually set off one another; so they all conspire to adorn the Earth: Insomuch that to suppose it of the prediluvian Form,

Form, would be rather to detract from its *measures*, than improve them. Yea, it would be in a manner to make it no Earth, or at least not so perfect a one as it is. For as we can have no Camels without Bunches; nor Mules without Hairs; nor Fowls without Feathers; or if we could, they would be but the more imperfect: so were the Earth abstracted from its aforesaid appendages, however it might have the more uniformity in it, yet as an Earth, it would have the less *comeliness*.

Somewhat to inforce this, Were a Man to contrive a Prospect for himself, we may be sure he would not have it all of a piece, or alike throughout: but would have it cast into Swamps and Hillocks, Bottoms and Gibbosities, Evennesses and Asperities; yea, into Seas and Ilets, and Rocks, if it could be; and so it would be an Image, not of the primitive, but present Earth. A petty Argument to prove, that there is something of perfection, or at least of pleasingness, in this Earth's disorder (if we may call it so) and that it is fitter to gratifie its principal Inhabitants (and so far) better in it self, than if it had been regular and undiversified. And the truth is, several of those appearances, which we are apt to call *rude*, *confused* and *uncouth*; and to count but Blemishes, Scars and Deformities; are commonly so well placed and suted to one another, as to become very taking in artificial Draughts, and a kind of natural Landskips.

And however the *Theorist* does sometimes disparage the Mountainous parts of the Earth at such a rate, as if they had been wholly unworthy of the care of Nature, and she had scorned to put her hand to the work of their Formation (and indeed his *Hypothesis* makes them nothing but *ruines*:) yet another while, when the ingenious Man is pleased to turn the stream of his Eloquence the contrary way; he represents them (though certainly the most horrid visible pieces of Nature) as exceeding *(a)* grateful to Beholders. Yea, he makes this very Earth of ours, and that in the hideously amazing and gastly Cragginess of its Mountains, to afford more delights to contemplative Minds, than ever the *Roman* or *Grecian* Theaters did, or those Sports wherewith they entertained Spectators. So he expresseth himself in the *Latin Theory* (Pag. *(b)* 89, 90.) And at the same time we find him transported as it were into a pleasing rapture or pang of Admiration, through the singular content and satisfaction he found, from the prospect and consideration of what we speak of. And truly that roughness, brokenness, and multiform confusion in the surface of the Earth; which to the inadvertent may seem to be nothing but inelegancies or frightful Disfigurements; to thinking Men, will appear to be as the Tornings, and Carvings, and ornamental Sculptures; that make up the Lineaments and Features of Nature, not to say her Braveries. Nor need we wonder that the *Theorist* should be so mightily pleased and raised, by the sight and contemplation of these things; for though some would take them for flaws and botches, and the fag ends of Nature; yet in them, a quick and piercing Eye can easily discern, not only *her* pretty dexterous Mechanisms; but the marvellous and adoreable Skill

(a) *Et quanquam revera semper horreant loca montana, & tesqua, ut jam diximus; non deest tamen in tanta varietate, quod recreet animum: atque sæpè loci ipsius insolentia & spectaculorum novitas delectat magis, quam venustas in rebus notis & communibus. Jucundum est ex profundâ valle prominentia montium supercilia, & impendentes moles suspicere, &c.*

(b) *Si quod verò Natura nobis dedit spectaculum in hâc tellure, verè gratum, & philosopho dignum, id semel mihi contigisse arbitror, &c. Hoc theatrum ego facilè prætulerim* Romanis *cunctis,* Græcisve, *atque id quod natura hic spectandum exhibet, scenicis ludis omnibus, aut amphitheatri certaminibus.*

Skill of her *Maker*, most rarely expressed. And therefore the inspired *Psalmist*, meditating upon the Earth in its *present Form*; and particularly revolving in his Holy Thoughts, the *Mountains*, the *high Hills*, the *Rocks*, and the *great and wide Sea*; was so taken with them, that he could not but think they had GOD for the cause or Author of them. And accordingly he declared and proclaimed the worst of them, not only to be produced by him; but to be the product of his infinite Wisdom. *O LORD in* wisdom *hast thou made them* all, Psal. 104. 24. And when the Divine Wisdom brought forth the Earth and these pieces of it, and ordered them into their present places and postures; and so admirably well, as that the Psalmist, directed by the Heavenly SPIRIT, could not chuse but celebrate the Production and disposition of them: has not *this* Earth as much to shew for its being made by *Rule and Measure*, as *another* of a pretended different Form, could have had? especially when it must all over have been but one vast Plain.

And then, in the Second place, this Form of the Earth is most *Useful* likewise. It appears to be so in sundry respects, and very considerable ones.

For now a *great part* of Mankind live by the Seas, either in way of Traffick or Navigation: not to say that *all* are some way or other the better for them. But in the First World, says the *Theory*, there was no Sea.

Mountains also now are most eminently serviceable; That is to say, in Bounding Nations; in Dividing Kingdoms; in Deriving Rivers; in Yielding Minerals; and in breeding and harbouring innumerable wild Creatures. I might also add, in contributing somewhat towards enlarging the Earth,

 and

and inabling it, in some Countries, to sustain its Inhabitants. Thus it is alledged as one Reason why *Palestine* could maintain so many of old: that *the Country was rising and falling into Hills and Vales, whereby ground was gained, and so the Land was far roomthier*, to use my * Author's Phrase. And indeed that there were store of Hills in *Judea*, and very fruitful ones, is insinuated by the Royal Prophet, where he calls upon Men to give praise to GOD, for † *making Grass to grow upon the Mountains*. But in the first Earth there were no Mountains neither.

* Fuller's *Holy War*, l.1.c.18.

† Psal. 147. 8.

Lastly, The Earth in its present Form and State, is attended with Rains and seasonable Showres. Whereas in its other Figure and capacity, it must have been all over cut into Rills and Aqueducts, for the Watring of Mens Grounds; and their trouble in doing it would have been endless and unspeakable, because it must generally have been done by hand. What Tongue can express the toil they must have had, in a manual watring of Fields, Woods, Groves, Orchards, *&c.* and in slicing a great part of the Earth in pieces, thus to moisten and cultivate the rest? But now kind Nature saves them that labour, while Clouds do the work effectually for them. For they filling their Buckets by the help of the Sun, and then emptying the same to the best advantage; excuse them from the drudgery, by taking it upon themselves. And that these Rules whereby we measure the *Usefulness* of this Earth, and shew it to be more excellent than that of the *Theory*; are the most true and proper Rules: is manifest from GOD's making use of the same in a Case not unlike: For he comparing *Egypt* and *Palestine*, prefers the latter before the former; because

becauſe in *Egypt* the Seed ſown was *watered with the foot as a garden of herbs*; but *Paleſtine* was a *land of hills and valleys, and drank water of the rain of heaven*, Deut. 11. 10, 11.

So that if an Earth moſt *comely* and decent in it ſelf, and alſo moſt *Uſeful* and convenient for Men; may moſt properly be ſaid, to be *laid in meaſures*, and to have had *the line ſtretched upon it*, or the Rule applied to it (as queſtionleſs it may) than the *preſent* Form of the Earth, may challenge this Text more juſtly to it ſelf, than the *other* could do, had it ever been.

And however the Architecture of *that*, is preſumed to ſurpaſs the Architecture of *this*; yet one thing may here be remarked concerning it, That the Holy Man's Language does but *indifferently* ſute it. For to talk of *Foundations*, in ſuch a *Circle*; or of a *Corner-ſtone*, in ſuch a ſpherical *Arch*, as the primitive Earth is conceived to be; ſounds but harſhly.

The Sixth Place conſiſts of the 8, 9, 10, and 11*th* Verſes of the ſame Chapter, where GOD continues his Interrogatories thus, *Or who ſhut up the Sea with Doors, when it brake forth as if it had iſſued out of a Womb? When I made the Cloud the garment thereof, and thick darkneſs a ſwadling band for it, and brake up for it my decreed place, and ſet bars and doors. And ſaid, Hitherto ſhalt thou come but no farther, and here ſhall thy proud waves be ſtaied.* Which Period the *Theory* would have to be underſtood, of the breaking forth of the Sea, at the opening of the Abyſs; but the Context allows it not. For that plainly ſignifies, that what the Sea is here ſaid to do, and what is ſaid to be done to that, was tranſacted in in the beginning; when the Foundations of the Earth were fastened, and the corner-stone thereof was laid, and the morning Stars sang, *&c.* And therefore when the *Theory* would put a difference (in respect of time) betwixt the foregoing (4, 5, and *6th*) Verses, and those last set down; so as to make the Questions in the former Verses, *proceed upon the Form and construction of the first Earth*; and those in the latter, *upon the demolition of that Earth, the opening of the Abyss, and the present state of both*: what it says, is *gratis dictum*, and the distinction groundless. Yea, it seems not only to be applied without grounds, but with force and violence; for the Context intimates no such matter, but rather the contrary. It runs on in a direct *series* of Queries, without giving the least hint, that any of the Particulars touching which they are made, were of later date than others. And that the first *set* of them, relate to things as ancient as the Primitive Earth's Production, the *Theory* owns; and therefore why should not the other too?

To which add, when the Sea brake forth at the time of the disruption, it could not be said to issue as out of a Womb, so properly, as out of its House (where it had dwelt above Sixteen hundred Years;) for a Womb is the place where a thing is conceived and brought into being, which before *was* not. But these Waters were preexistent to the inclosure of the Abyss, the Womb which held them; yea against the order of Nature, they were contributive to the being of it, as they were the *basis* whereon the First Earth was built. So that the place of the Abyss falls in but ill with the notion of a *Womb*,

Vomb, in reference to these Waters; And consequently they could as ill be said to issue from thence
s out of a Womb. And then the *Darkness* at the
Disruption was not so *thick*, nor so much a *garment*
or *swadling band* to the Sea, as darkness was at the
Creation. Yea, the truth is, it could then be no
garment or swadling band at all for the *Sea*, but
only for the *Flood*. For by that time the tumul-
tuary Waters of the Deluge, were quietly retired
into the *decreed place*, and became a *Sea*; the Sky
was cleared up, and the darkness gone. Nor could
t so properly be said to be *shut up with Doors*,
and to have *Bars set upon it* then, as to be in-
franchized or set at liberty. For those Doors and
Bars which shut it up, and made it fast in a closer
state before the Disruption, were then all broken
down and thrown open for ever, and it was put
into a condition of far greater freedom than it formerly had; its present settlement, being perfectly a
state of enlargement to it.

But now turn the words to the sense of the
Old *Hypothesis*, and (besides that they keep time
exactly with the Context) how patly do they fall
n with it? For when on the First Day, GOD
(together with the Earth) made the Water of the
Sea; as *it brake forth* into being, *as if it had issued
out of a Womb* indeed, because it just then gushed
out of the *Womb* of *nothing*, into Existence: and
as he then *made the Cloud the garment thereof, and
thick darkness a swadling band for it* in a fuller
sense; for *darkness was then upon the face of the deep*,
Gen. 1. 1. and that darkness for certain most *thick*,
there being then neither Sun nor Light: so on the
Third Day, when he brake up Chanels for it, he
might

might well call them *His decreed place*, and declare that he had beset it with *Bars and Doors* ; because by his command the Waters were gathered off the surface of the Earth, where was their first and natural situation, and shut up in such Receptacles, and with such a confinement, as they would never have withdrawn into of themselves ; but would always have remained in their original diffusion over the whole Terrestrial Globe. And that this shutting up of the Sea in its *decreed place*, was a thing done in the beginning, and not at the time of the Flood ; is evident, *Prov.* 8. 29. where GOD's giving his Decree to the Sea that it should not pass his commandment ; and his appointing the foundations of the Earth : are made to be Synchronals.

But from the last Verse of the Quotation, *Hitherto shalt thou come and no farther, and here shall thy proud waves be stayed* ; an objection is raised against the usual exposition of the Place. For that *sentence shews* (saith * the *Theory*) *that it cannot be understood of the first disposition of the Waters as they were before the Flood, for their proud waves broke those bounds whatsoever they were, when they overflowed the Earth in the Deluge.* I answer, If they did so, yet that argues not but the words may speak the disposition of the Waters before the Flood, according to the common interpretation of them ; for that Inundation was by GOD's special appointment. And when he assigned to the Waters the place of their abode, he did not intend to fortifie them in it against his own Omnipotence, or to devest himself of his Sovereign Prerogative of calling them forth when he pleased. And when they passed the bounds he set them, so long as they did it not by any force of their

* Page 89.

their own, but meerly by his powerful order or providential act; this their Eruption and spreading Overflow, cannot be lookt upon as a breach of that Law, or those Limits he prescribed them. It was only the marvellous effect of an extraordinary Cause; and a particular Exception of GOD's own making, to the general and standing Rule of his Providence. Just as *Enoch*'s, or *Elijah*'s Translation was, to the universal and irrevocable Sentence of Death. That may be one answer in defence of the ancient *Hypothesis*.

But then, to the *Theorist*, I may give in this for another: The proud Waves of the Sea did never pass their bounds to make the Deluge. The great Deep, or the Fountains then broken up, had no relation to the Sea; I confess this implies that the Flood is to be explained by a new *Hypothesis*; but if we can but bring in such a one, as may be as justifiable as the *Theory*'s is (which we shall endeavour to do) we need not concern our selves farther about it.

The last place is *Prov*. 8. 27, 28. *When he prepared the Heavens, I was there; when he set a compass upon the face of the Deep: when he established the Clouds above, when he strengthned the fountains of the Abyss.* Whence is inferred, * *So there was in the beginning of the World, a Sphere, Orb, or Arch, set round the Abyss*, which is presumed to be no other than the first *habitable Earth*.

* Theor. p. 50.

But this is a sense far fetcht to serve the turn of an *Hypothesis*, when there is a nearer at hand will do much better. For by the *Compass* set upon the face of the Depth, is meant no more than those bounds wherewith GOD encompassed (not the *Theory*'s

Abyss, but) the *open Waters*. The HOLY GHOST (who is the best Interpreter of his own Writings) expounds it so by a paralled Text in *Job*, *He hath compassed the waters with bounds*, chap. 26. ver. 10. Take it in the Original, and it speaks out *Solomon*'s meaning to the full חק חג על פני מים, *terminum circinavit super faciem aquarum*; *With a pair of Compasses he set a boundary upon the face of the Waters.* Not upon the face of the *Deep*; so it might have been catcht at, and construed an *Arch* upon the *inclosed* Abyss: but upon the face of the *Waters*. And this *Compass* was extant in that state of Nature, where were *Thunders*, and *Waters in thick Clouds* (as the Context shews) neither of which *Phænomena's* could be contemporary with that Arch or Orb which the *Theory* contends for. And then it was to last *until day and night come to an end*. So that if *Solomon*'s meaning be the same with *Job*'s; the *Compass* he mentions as set upon the face of the Deep, must be standing still. And so it cannot be that Arch which the *Theory* would perswade it was, because that was down long before *Job*'s or *Solomon*'s time. And yet that these two great Men (both *Kings*, as some think) did intend the same thing, the *Theory* * acknowledgeth. And that חק does here signifie a *Boundary*, may well be inferred from what follows in the next Verse, *when he gave to the Sea*, חקו, *his decree*; which the *Targum* renders, תחומה, *his boundary*. Or if the Phrases used by *Solomon* and *Job* sute not so exactly with the Waters as encompassed with *Earthly* bounds; yet they are very applicable to them, as they are encompassed with the *firmament* of Heaven. For *that* is set as a Sphere or Orb, as an Arch or Circle

* *Idem velle hæc duo loca* Solomonis & Jobi, *in dubium est*, p. 256.

Circle *upon the face of the Deep*; and ſhall continue עד־תכלית אור עם־חשך *until the conſumption of light with darkneſs*, according to the Holy Man's expreſſion. And this the old *Chaldee* Tranſlation falls in with, while it ſays, GOD ſet the *firmament* upon the Waters. And ſo does *Eugubinus*, who affirms, That the place in *Job*, is to be underſtood *de orbe Cæleſti*, *of an heavenly orb*, † as the *Theory* has noted to our hand; though that he did it, || *parùm philoſophicè*, we have little reaſon to believe, when we read of, חוג שמים, the *Orb* or *Circle*, the *Sphere*, or *Compaſs of Heaven*, Job 22. 14.

† Pag. 257. Lat.
|| Ibid.

And then by GOD's *ſtrengthening the fountains of the Deep*, is meant his making the Earth ſo compact and ſolid; as that the Springs and Rivers derived from the Sea, ſhould not ordinarily waſh it down, and ſo obſtruct and dam up themſelves. But how on the other ſide, an Arch built over the *Moſaical* Abyſs, ſhould any way *ſtrengthen the Fountains of that*, when not ſo much as one Spring, or River, or fountain *in ſpecie*, did ever flow out of it during its incloſure; is not ſo eaſie to apprehend.

4. Such are the Scripture-Proofs of the aforeſaid Form of the Antediluvian Earth. To take them off, I might oppoſe them by many other Texts: I mean ſuch as are charged with counter-Metaphors; with ſuch Allegorical or alluſive terms, as carry a ſenſe in them, not only different *from* what is ſuggeſted in the forementioned Allegations; but inconſiſtent *with* it, and repugnant *to* it. I will inſtance but in one; *Who ſhaketh the Earth out of her place, and the Pillars thereof tremble*, Job 9. 6. So that the Earth, which is one

while ſaid to be founded upon the Seas, and eſtabliſhed upon the Floods:and another while to be ſtretched over empty places, and hanged upon nothing: and anon (according to the *Theory*) to be a Sphere, or Circle, or an independent Orb or Arch; is ſaid at laſt to be built upon Pillars. Whence it is manifeſt that the Citations above, are but Tropical or Figurative Schemes of-ſpeech; and ſo wide and indeterminate, that nothing of ſtrict and particular ſignification or certainty, is to be lookt for in them, or concluded from them. To do that (though I will not ſay it is to trifle with Scripture) is to make it ſpeak what it never meant.

It is ſaid of GOD, in the cited Text, That *he ſhaketh the earth out of her place.* Which had it been hit upon, and that way applied, would have been as notable an evidence for the Earth's changing her ſituation in the time of the Flood (by ſome terrible concuſſion happening to her in her Diſſolution) as any the *Theory* has brought to other
* Pſal. 104. 5. purpoſes. And yet we read in * the *Pſalms* that GOD *founded the Earth,* על מכוניה. *upon its baſis, that it ſhould not be removed for ever.* Though at the ſame time we are told again, *The earth is diſſolved,* Pſal. 75. 3. quite down as it were, and all in ruines (which might have been a Proof of its Diſſolution at the Deluge) even then when it was impoſſible alſo that it ſhould be ſo, becauſe GOD upheld it; for it follows immediately, *I bear up the Pillars of* it. Moſt plain Demonſtration how little of Argument, as to the matter in hand, can be drawn from ſuch σχηματολογίαι, or Tropological forms of ſpeech as theſe, which frequently occur in the ſacred Volume, eſpecially in the *Poetic* Books thereof.

5. Had

5. Had learned *Tycho* but minded this, and rightly considered how the HOLY GHOST does all-a-long deliver himself in *Figurative* Expressions touching the Earth; he needed not to have scrupled the *Copernican* System of the World, and (falling off from the old *Pythagoric* Hypothesis) have erected a new one of his own, more intricate and less tenable; in tenderness to the Sacred Writings. For † *Gassendus* gives *that* in, as one of *Brahe's* Objections against *Copernicus's* way, (and as one reason for his inventing and setting up his *own*) *quod Sacris adversetur Literis aliquoties ipsius Terræ stabilitatem confirmantibus.*

† *Inst. Astron.* lib. 3. cap. 13.

CHAP.

CHAP. VIII.

1. *A* continual Æquinox *before the Flood, by virtue of the Earth's* Position, *improbable.* 2. *For then that* Position *would have* remained *still, or the* Change *thereof would have been more* fully *upon* Record. 3. Scripture *does not* favour *this* Æquinox, *but rather* discountenance *it.* 4. *It would have kept one* half *of the Earth* unpeopled. 5. *And have hindred the* Rains *at the time of the* Flood. 6. *The* Doctrine of the Æquinox *is against the* Judgment *of the* Learned. 7. *The* Authorities *alledged for the* Right Situation *of the Earth, upon which the Æquinox depends,* Insufficient *to prove it.* 8. Two Queries *propounded relating to the* Æquinox.

1. WE are now (from its *form*) come to the *First Property* of the Antediluvian Earth, namely, a *Perpetual Æquinox* by reason of its right situation to the Sun. By which is meant that the *Axis* of the Earth was always kept in a Parallelism to that of the Ecliptic (as now it is to that of the Æquator.) So that in her Annual motion about the Sun, she was carried directly under the Æquinoctial, without any manner of Obliquity in her site, or declination towards either of the *Tropics* in her Course; and therefore could never cut the Æquinoctial, by passing (as now she is presumed to do) from one *Tropic* to the other. The effects of which her regular position and motion, during the continuance of the same; were an even and unvaried Temperature of the Air; a constant Spring and unwearied

unwearied fruitfulneſs in the Earth; and an uninterrupted tenour in the interchanging viciſſitude of Days and Nights, they being ever of a length. But ſuch a direct ſituation and courſe of the Earth, is a thing very unlikely to have been.

2. One reaſon is this; becauſe then the ſame would have remained until now; or elſe in the World there would have been found a more full account of the Change thereof. For put caſe the Earth did ſhift her poſture, and alſo her Circuit about the Sun, in which ſhe perſiſted till the Deluge. Is it not odd and monſtrous ſtrange, that we ſhould have no more to ſhew for this? and that no better footſteps of its remembrance ſhould be ſeen?

Whenever this Change befell the Earth, as to its ſite and yearly progreſs; it muſt needs have been attended with a notable alteration in the *Tempeſtival* (to ſay nothing of the *Aſtronomical*) face of things. And when they who had eſcaped the common Shipwrack, came forth of the Ark; and beheld how the days did lengthen and ſhorten; and how the Year ran through ſuch variety of Seaſons: and withal felt their Bodies ſo differently affected, being one while parched with Summers heat, and another while pinched with Winters cold; contrary to what they had ever been before: they could not ſurely but relate this ſurprizing Novelty to after-Generations (as a thing moſt wonderful) and they tell it to others, and they again to others: and ſo ſome dark account of it at leaſt, would ſomewhere have been met with, amongſt the broken Records and Monuments of Antiquity, more than now appears. Not to add, That where Providence does

does make so considerable Changes, and of so near and important concern to Men; it does usually register them, and give fair notice thereof to Posterity. Thus in a National concern of the *Jews*, there being but one day made *longer* than any had been, we find it distinctly recorded in * *Joshua*, and afterwards confirmed by the Prophet † *Habakkuk*; And therefore had this Alteration been real, methinks it should have been more fully recorded too; especially when so fit an occasion of Chronicling it was offered of old, when the Story of the Flood was committed to writing.

* Chap. 10. Ver. 13, 14.
† Chap. 3. Ver. 11.

There once happened a notable Change in the Planet *Venus* (the *Theory* has remarkt it) remembred by *Castor*, and out of him by *Varro*, and out of him by St.* *Austin*. And this was said by *Adrastus Cyzicenus*, and *Dion Neapolites*, two noble Mathematicians, to fall out in the Reign of *Ogyges*. By whom if they meant *Ogyges Priscus*, who was *Noah*, the Date of the *Catastrophe* was about the time of the Deluge. But then if a single Change in one of the Heavenly Bodies was thus noted, and the notice of it so plainly transmitted to us at such a distance: then had a general Change at the same time befallen the Heavens, the whole aspectable Heavens, and the Earth at once; certainly we should have heard something more concerning it than now we do, from the famous Ancients.

* *De Civit. Dei, l.* 21, *c.* 8.

Though when that Planet did (according to the Historian) change her *colour*, *magnitude*, *figure*, and *course*; we need not impute this (as the *Theory* does) to her then present *dissolution*: but rather to the disposition and temperament of the Air, which perhaps will be able to solve all the *Phænomena*'s. For grant but *that* to have been full of moist

Vapours,

Vapours, and of a constitution so watry, as it never was before nor since (which it might very well be, and could scarce be otherwise about the time of the Deluge) *Venus*, by unusual refraction of her Beams would easily put on a different *hue*, and larger *Phase* than she used to wear. The same Air also might alter her *shape*, while the humid *medium*, performing the part of a *Telescope*, truly represented her *gibbous*, *corniculate*, or the like. And then it might put her *Course* into seeming Disorder too. For the Air above being unequally thick, and subject at times to uneven agitations; as it chanced to be variously driven or moved, might fling the Planet into unsteddiness as to appearance, or into a kind of fluctuating or salient Motion in the Eyes of Spectators. And so it might seem to be (as *Marcus Varro* terms it) *mirabile portentum*, a wonderful monstrous thing. But that the whole mutation or disorder which happened to this Planet, is no good Argument of its being just then dissolved; is evident from this Passage in the Story which speaks it to have been but *temporary*: *Quod factum ita neque antea, neque postea fit*, *It was never so before nor after.* And therefore still the more probable it is, that the Air which was then so out of order too as it never was before or after, might be the cause of all. And why these effects should be visible only in this Planet, there is more to be said than needs be here inserted. Let me but hint, that if the Moon were then in Conjunction, or near it; *Venus* was the brightest Luminary that shone by night, and so the fittest for these *Phænomenas* to shew forth themselves in; especially she being subject to *increase* and *decrease*.

But to return, Though *Moses* did not commemorate this mighty Change, when he had so fair an occasion of doing it, in the Story of the Flood; yet had it really happened to the World, it could not have slipt so perfectly out of memory, as it has done. For at the time it fell out, there wanted not *one* at least, who was very well able to remark it; and to have given occasion (by passing his Observations concerning it to others) to a lasting traditional remembrance of it; I mean *Noah*. And that he was qualified for this, we need not doubt, if what * some report be true; namely, That the famous *Atlas* (who for his Skill in Astrology is fabled to support the Heavens with his Shoulders) was *Enoch*. For if *he* were so eminent in that sort of Learning, *Noah* might be rarely versed in the same; at least he must have been so competently instructed in it, as to have been capable of leaving a most clear account behind him of this marvellous alteration, if it had happened in his time. For Books written by *Enoch*, are reported to have been preserved in the Ark. And *Origen* affirms, That part of these Books, containing the course of the Stars, their names, *&c.* were found in *Arabia Felix*. And *Tertullian* avers that he had seen and perused many Pages of them. And Sir *Walter Raleigh* (no bad Historian) is so far from condemning or suspecting the thing, that he rather vindicates it. Though it is not to be doubted, but into them at length many extravagancies might be inserted. Now if these Books treated of *Astronomy*, as *Origen* says they did; *Noah* could not chuse but derive good Skill in that Science, from them. And so (by the way) it will be easie to conceive how *Abraham* came to such perfection in it, as to impart it to the

* *Euseb. Alex. Polyhist.*

the *Chaldeans, Ægyptians,* &c. as by * *Josephus* he is said to have done. For he being near Sixty Years old when *Noah* died, by living and converſing with him ſo long, he might gain ſo much knowledge in Aſtronomical matters, as to be able to inſtruct thoſe Nations in them. Eſpecially if he addicted himſelf ſo much to the ſtudy of Aſtronomy, as that *that* gave him his name *Abram*; as the Knowledge of GOD cauſed *Alpha* to be put into it, and turned it into *Abraham.* For ſo a Learned * Man has given us to underſtand; That *he purſuing the high Philoſophy of things that happened in the Air, and of thoſe aloft that move in the Heavens, was cal'ed* Abram, *which is interpreted,* **Sublime Father.** *But afterwards* ——— *he takes* Alpha *into it, the knowledge of the one and only* GOD, *and is called* Abraham.

* *Antiq. lib.* 1. *cap.* 8.

* Οὗτος, τὸ μετάρσιον τῶν κατὰ τὸν ἀέρα συμβαινόντων, καὶ τὴν μετέωρον τῶν κατὰ τὸν οὐρανὸν κινουμένων φιλοσοφίαν μετιών, Ἀβρὰμ ἐκαλεῖτο. ὃ μεθερμηνεύεται, πατὴρ μετέωρος. ὕστερον δὲ —— περιλαμβάνει τὸ ἄλφα, τὴν γνῶσιν τοῦ ἑνὸς καὶ μόνου Θεοῦ, καὶ λέγεται Ἀβραάμ. *Clem. Al. Strom. l.* 5. *p.* 549.

Indeed it is not to be doubted but a great deal of the ancient Learning is loſt, as † the *Theory* concludes. And he that obſerves what a multitude of Books are ſaid by *Laertius* to be written by *Xenocrates, Theophraſtus, Democritus,* and others; of which ſo few are now to be found, will eaſily believe it. But yet this will not ſatisfie as to the deep ſilence of Antiquity touching the *Æquinox* aſſerted, or the change thereof. For other *Theorems* or *Dogmats* (even far more remote from notice, and of a nature every whit as obſcure or inevident, though of late cleared up) have been plainly delivered by ſome Philoſophers or other; and ſafely handed

† Book 2. Chap. 9.

down to us, either in *their*, or in *other Men's* Writings.

Thus *Pythagoras*, as *Laertius* relates, taught the Earth to be περιοικουμένην, *inhabited round about*: εἶναι δὲ καὶ Ἀντίποδας, and *that there were Antipodes*, to whom, τὰ ἡμῖν κάτω, ἐκείνοις ἄνω, *the things under us, were above to them*. A Doctrine heretofore as little approved, as believed; and so ill thought of, that the asserting it has cost some Men dear. To which add what *Plutarch* in the Life of *Numa* remembers; That the *Pythagoreans thought the Earth* οὔτε ἀκίνητον, οὔτε ἐν μέσῳ τῆς περιφορᾶς, *to be neither immovable, nor placed in the midst of the vortex*, or center of the turning Region; ἀλλὰ κύκλῳ περὶ τὸ πῦρ αἰωρουμένην, *but to be hung up in a circle* running *about the fire*, that is, *the Sun*. (The very *Hypothesis* revived by *Copernicus*, and improved by *Des-Cartes*.) And to typifie the Sun's being seated in the center of that Heaven in which he shines; the same *Numa*, says *Plutarch*, built the Temple of *Vesta* in a circular form, and placed τὸ ἄσβεστον, *the fire never to go out*, in the middle of it. *Leucippus* also (as we find in the aforesaid *Laertius*) affirmed, τὴν γῆν ὀχεῖσθαι, περὶ τὸ μέσον δινουμένην, *that the Earth was carried round*, or rolled about *upon its own* axis. From whom likewise we learn, That *Anaxagoras* was of opinion, ἀστραπὰς ἐκ τρίψεων νεφῶν, *that Lightnings were caused by collision of Clouds*, as also τὴν σελήνην οἰκήσεις ἔχειν, ἀλλὰ καὶ λόφους, καὶ φάραγγας, *that the Moon was habitable, and full of Hills and Dales*. As if *Galileo*'s Glass had been an old Invention, and this Philosopher had known as much of the Moon above Twenty Centuries ago; as *he* discovered of late, and has given the World an account of in his *Sidereus Nuncius*. *Heraclides* also, as * *Plutarch* reports, believed

* *De Placit. Phil. lib.* 2. *cap.* 13.

lieved ἕκαστον τῶν ἀστέρων κόσμον ὑπάρχειν, that *every Star was a World, comprehending in a vast æthereal ſpace, an Earth*, &c. And ſo thought *Orpheus.*

Now when theſe and the like pieces of Philoſophic Learning, were preſerved in the midſt of that Shipwrack which it ſuffered; it is ſtrange that the Dint of Fate, ſhould fall ſo heavily on this ſingle Notion of a *Perpetual Æquinox*, as to ſink it down to the bottom of Oblivion, and leave us not ſo much as *one clear* Aſſertion of its exiſtence or expiration. For however the *Latine Theory* tells us of * *teſtimonia ſatis illuſtria, teſtimonies clear enough* to evidence the *right Poſition* of the Earth, and conſequently the *Æquinox* depending upon it: yet when we come to examine them, we ſhall find they are but blind and cloudy things, and without all ſolid reaſon for their Foundations. So that upon the whole matter, the Aſſertions concerning the *things aforeſaid*, which are *ſupra nos*, more out of the way (except the *Antipodes*) and leſs ſubject to Obſervation, than the Situation of the Earth, and the Æquinox attending it, and the change of both; are more expreſs and rational, than any of the *Teſtimonies* concerning *Theſe*: Which is ſomewhat ſtrange, I ſay, theſe being *Phænomenas* which of old fell under common notice in way of Experience; whereas the other were never ſo obvious and tried. And the more ſtrange will it ſeem yet, if what was hinted before, be duly conſidered; namely, That *Noah* might be well qualified to obſerve ſo great and remarkable things, and to recommend the Obſervations to his Poſterity.

* Pag. 291.

3. As

3. As for *Scripture*, it is so far from favouring this *Æquinox*, that it does rather discountenance it. And those words, *Gen.* 8. *ult. while the earth remaineth, seed time and harvest, and cold and heat, and summer and winter, and night and day shall not cease:* instead of any change in the frame of Nature (which the *Theory* would infer from them) intimate the contrary; that things still continued in their former State; and were not out of a more regular and uniform, *then* put into a new and less orderly course and posture, than they were in before. For,

First, The words seem to look so directly the other way, that they can scarce be made to cast an eye on such a sense, without violent Distortion of their natural Aspect. *Noah* was just now come out of the Ark; and having so dismal a prospect before him, so black and horrid and amazing a Spectacle, as the utter Destruction of all Mankind, excepting himself and seven more: this might very well damp him extreamly, and fill him with melancholy and sad Dejection. And then the dreadful apprehensions of what might yet be behind, or happen again afterwards of the like nature; might startle him exceedingly, and fright him into farther Consternation. Now to support the good Man under the weight of this double terror and solicitude, or to take off its heavy pressure; GOD here passeth a solemn Promise, that no such Flood should ever drown the Earth any more. And then in Confirmation of this Promise, adds, That the yearly *Seasons* should never thenceforward be interrupted; which they certainly must be, in case of such another universal Deluge. That this was the occasion and full scope of these words,

words, * *Josephus* attests with advantage on our side; For he says, That *Noah* (upon his coming out of the Ark) *fearing lest the Earth should every year be overflowed, offered burnt sacrifice to GOD, beseeching him that hereafter he would entertain the* ancient *order*, &c. To which request of his, what more gracious or satisfactory answer could be returned, than in the words recited? Where GOD condescends to give him assurance of what he desired, by ingaging, That *while the earth remaineth, seed time and harvest, and cold and heat, and summer and winter, and night and day shall not cease.* Where *Summer* and *Winter* are mentioned, as things well known to the Patriarch, and he makes no enquiry into the *meaning* of them, as having been familiarly acquainted with them.

* *Antiq. l.* 1. c. 4.

Secondly, GOD here promiseth to *Noah*, in behalf of Mankind, That there should be *Day* and *Night*, as well as *Summer* and *Winter*; yet Day and Night were certainly before the Flood; and if the promise of *their* continuance does not hinder but they were before; so it argues not but that *Summer* and *Winter* were so too. Yea, since *Summer* and *Winter* are here settled upon the new or recovering World, in conjunction with *Day* and *Night*, which had their alternate beings ever since the Creation; it is a good evidence that these *Seasons* had the same. And the reason why both were now ensured, is, because both were intermitted; the Rule of Day and Night, having been broken for a while, by continual darkness; as well as the Regularity of the Seasons (for that fatal Year) by the prevailing Waters. To which add,

Thirdly,

† Gen. 1. 14. Thirdly, That † the *Lights in the Firmament of Heaven*, at the same time that they were appointed to *divide the day from the night*, were moreover appointed for the *Seasons* of the Year; for so מועדים there signifies. And therefore those Birds that come in the *Spring* and go away in the *Autumn*; and are in one place in the Summer, and in another in the Winter; are said to *know their appointed times*, or the *Seasons* of the Year, *Jer.* 8. 7. and the Prophet expresseth them by the same * word that *Moses* did. But,

* מועדיה.

Fourthly, There is annother thing, wherein Scripture checks with this *Æquinox*; and that is the effect of the Divine Malediction denounced against the Earth. Upon Man's rebellious defection or Apostasie from G O D, he cursed the Ground for his sake, *Gen.* 3. 17. Whereupon it became naturally barren of good things necessary to Life; and fruitful in useless and offensive Products. But in case there were such an Æquinox, it will be hard to conceive how this should be; for that Æquinox would have kept the Heavens in a standing unvaried posture; and the stability and unchanged influence of the Heavens, would have continued the Air in the same benign Temperature. And the Air being still, and warm, and balmy; that rich and fat Earth would have been flourishing and fruitful, pleasant and *Paradisiacal* (as the *Theory* supposes it) a long time after *Adam* fell. So that where could be barrenness? Or how did the Curse of GOD take place? To say the Earth grew dry and barren at last, for some ages before the Flood, would be no answer, or at least no satisfactory one. For besides that the heavy Curse was presently to fall as a Punishment upon *Adam*; so late a barrenness would

would have been the effect of time and nature; the unctuous juices of the primigenial Soil, which made it a great while ſo vital and vegetative, being at length exhauſted. And therefore this barrenneſs could not be imputed to the Curſe of GOD, becauſe it would certainly have come on in the meer courſe of things, though Man had perſiſted in his original purity, and had kept the Crown of integrity always upon his head.

Laſtly, There is a Paſſage in the Holy Writings, which ſeems to evince, That the *Air* in Paradiſe, had an *Intemperature* ſent into it (perhaps the fruit of the Curſe now mentioned) about the time that our Firſt Parents ſinned. And this again implys, That there was no ſuch Æquinox. The Paſſage relates to our firſt Parents, and occurs, *Gen.* 3. 7. Where it is ſaid of them, That *they ſewed fig-leaves together* (or * *fitted* them together, as the *Syriac* reads it) *and made themſelves,* חגרת, *things to gird about them.* Now why did they do this? It is commonly ſaid, *That they might cover their nakedneſs, whereof they were aſhamed.* But this ſeems not to have been the reaſon, at leaſt not the whole reaſon of the thing. For firſt, Scripture ſays nothing of it expreſly; That does not declare that they did thus to hide their ſhame. Secondly, What ſhame need there have been upon account of nakedneſs betwixt Husband and Wife, when there were no other People in the World? Thirdly, While they ſtood, it was ſaid of them, That *they were both naked, the man and his wife, and were not aſhamed,* Gen. 2. *ult.* And ſurely when they were *innocent,* they ſhould have been moſt *modeſt*; and their modeſty

* So תפר properly ſignifies, *Job* 16. 15. and *Ezek.* 13. 18. For who ever *ſewed* Sackcloth to his own Skin? or to other Mens Arm-holes?

should have made them most ashamed of their nakedness *then*, had there been shame in it. And therefore it is probable that the *Perizomata*, things to put about them, were made upon *another* score; namely, To defend them from the *intemperate Air* of the *Edenical* Regions. And this was as much as they at present could do for themselves. But then afterwards (which helps to confirm our sense) we find that *the* LORD GOD *made them coats of skins and cloathed them*, Gen. 3. 21. Which were to be a better defence still against the aforesaid Inconvenience. So *Lyra* concludes, That they were cloathed with Skins, *(a) because they wanted a covering against the* **Intemperature of the Air.** I confess he speaks of the Air in *that* place *ad quem erant ejiciendi, into which they were to be cast forth.* But let it be so: still it will fight as much against this Æquinox; and imply or infer what certainly overthrows it; that is, an Air *intemperate* in the habitable Regions of the first Earth.

(a) *Quod indigebant tegumento contra intemperiem aeris.*

And (by the way) let none wonder, That GOD, by his Angels, should stoop to so mean a work, as the cloathing of *Adam* and *Eve* with Skins. Let us but seriously think what disparaging things our REDEEMER JESUS the King of Glory, has done and suffered in his adorable Person, for us forlorn and most unworthy sinners; and we shall cease to marvel at this lesser condescension of the INFINITE MAJESTY, though it was exceeding great. Yet had it not been more upon the account of *warmth*, than *covering their nakedness*; such Coats need not have been made them: their own Fig-leaves would have been sufficient for that use. And thus Scripture does plainly discountenance

discountenance this Æquinox, rather than favour it in the least measure.

4. But farther yet. If the Earth always wheeled about the Sun, in a *Right Situation* to him; the Terrestrial Globe, in one Hemisphere of it, must have been *unpeopled*; because there could have been no easie Passage, no way of possible access to it. For grant *Adam* to have been planted on *either* side of the *Torrid Zone*; how should *he*, or *his*, have gone through it to the *other*? It would have been so terribly heated by the roasting Sun, that no Mortal could have travelled over it. Consider but the *breadth* of this Zone: According to the Ancients (who stretched it from one *Tropic*, to the other) it was about Seven and forty Degrees wide; that is, near Three thousand Miles. But yield it to have been but *half* so broad, and what Men could ever have marched over it? For as under their feet there would have been vehemently hot and scalding Sands: so the scorching fury of the glaring Sun, would have beat intolerably upon their Heads. And then what should have guided them through this burning Tract, where was nothing of Path, or Way-mark to be seen? Suppose they had the Direction of Stars by night; yet who, or what should have led their Caravans by day? And yet had they journied without sure conduct, whither might they have wandred? and to what length might they have spun out their rangeing Progress, at the shortest too long and tedious to be born? Especially if we consider, that in those their Travels, they could have met with no manner of shelter or refreshment: No, not so much as with a Grove, or a Tree; with a Lake, or a River; with one

poor Fountain or Spring of Water, or a single puff of fresh and cooling Air. And say the driest burningest part of this Zone, had not been above Five hundred Miles over; yet who durst have thought of venturing through it, as not knowing its extent? And who that had advanced a few Furlongs into it, could have been able to have gone forward, or to return alive? None will be surprized at this, that have a right Notion of the nature of this Region; or of the excessive degree of its raging heat. The *Theory* speaks it in these words, (which, all circumstances weighed, carry no *Hyperbole* in
* Pag. 257. them) * *It was a wall of fire indeed, or a Region of flame, which none could pass or subsist in, no more than in a Furnace.*

Now if *Adam* were seated at first in the Southern Hemisphere of the Earth, as the *Theory* holds, then how could *he*, or any of his Off-spring, have removed into this Northern one? there being such a fiery Partition betwixt them. Yet we are told of
† Pag. 371. Providence's † *transplanting* Adam *into this Hemisphere, after he had laid the Foundation of a World in the other.* But that *Adam* in any ordinary Providential way (and no extraordinary one is mentioned) should cross *a wall of fire* or *a Region of flame* (we know not how many hundred Miles broad) *which none could pass or subsist in, no more than in a Furnace*; may justly be concluded a thing impossible. And then equally impossible it was, that this Hemisphere of ours, should ever be peopled by *Adam* or his Progeny before the Flood.

To say that GOD led *Adam* through this Me-
* Theor. pag. 233. diterranean fiery Zone (*the * Barrier betwixt the two Hemispheres, which nothing could pass either way*) as soon as he had sinned; and so very timely, that it

it was not as yet grown hot and burning ; might be a uſeful ſuggeſtion in the caſe, were it not perfectly foreſtalled and quite ſhut out, by what was ſaid before ; namely, That *Adam* was not tranſplanted into *this* Hemiſphere, till *he had laid the Foundation of a World in the other.* Which ſuppoſe to have been done in *Twenty Years* time (as it could not well be done in leſs) yet in that interval, the fire would have been ſo kindled in the Torrid Zone, as to have made it too hot a Climate for him to have gone through.

If in this our Land we have no Rain for eight or ten weeks together in a Summer, we ſee how lamentably the Ground is ſcorched, and how the ſurface of it is turned as it were into a meer Turf: and yet all this while the Sun is not perpendicular to us, by two or three thouſand Miles. But how inconceivably hot then muſt the middle circumference of the Firſt Earth have been, ſuppoſing it ſubject to his perpendicular Beams, not only for ten weeks, but twenty years together: and no one Cloud to have overſhadowed it, and no drop of Rain to have fallen upon it, all that while?

It is ſaid to have been the Opinion of *Athanaſius*, and *Ephrem Syrus*, That Paradiſe, into which *Adam* was put, lay beyond the Ocean : and that he wading through it, made towards the Country where he was formed; and at length dying there, was buried in *Mount Calvary*. Upon what good grounds this conceit was built, I know not: but by no means can it eſcape the Cenſure of abſurdity. Yet the vaſt Ocean it ſelf might as well be fordable to the firſt Father of Mankind; as this glowing Zone, paſſable. And therefore the difficulty of getting through that Ocean, was one thing

thing that induced St. *Austin* to follow *Lactantius*, and the Ancients generally, in denying *Antipodes*. For in their Judgment an immense Ocean begirt the Earth (after the manner this Zone is supposed to have done) and parted our Northern from the Southern Hemisphere. For which reason, the good Father deeming it impossible, that the Ἀντίχθονες, or Inhabitants of that side of the Earth which is opposite to ours, should ever spring from the same Stock with us, and be of *Adam*'s Race; he fairly concluded that there were no such. * *It is too absurd to say that any Men could out of this, get into that part* of the Earth, *by sailing over the huge Ocean: as also* it would be to say, *That Mankind was founded there, of that first Man*, Adam. And t'erefore, by the way, how could St. *Austin* (if consistent with himself) place Paradise in the Anti-hemisphere, or Continent opposite to ours (as † the *Theory* understands he did) when he thus expresly declares it to be his Judgment, That Mankind was not propagated there, and could not be transported from hence, thither?

* *Nimis absurdum est, ut dicatur aliquos homines ex hac in illam partem Oceani immensitate trajectâ navigare, ac pervenire potuisse.* Civit. l. 16. cap. 9.

† Pag. 256.

5. Again, Had the Earth held such a *Right Situation* to the Sun; it would have put by the Rains, which helpt to raise the Flood. I confess it is granted, That at that time, * *the rains fell, forty days and forty nights together, and that throughout the face of the whole Earth.* And this is but a certain truth, and so a necessary concession. But then it is more than the *Hypothesis* can bear; which makes Rain impossible (while the first Earth stood) in any other place but the Frigid Zones. And therefore

* Theor. pag. 98.

therefore to admit ſuch *general* Rains, is to deſert or overthrow the *Hypotheſis*; and to ſuppoſe the Situation of the Earth changed, before it was ſo.

So incompatible were Rains to the firſt order or conſtitution of Nature, as fixed by the *Theory*; that a Particular Hydrography was calculated by it, to ſerve the prediluvian Age with Water. But then the ſame Syſtem or Frame of Nature, which rendred that World ſo impluvious all along, would have done ſo at the time of the Flood likewiſe. Yea, in that critical juncture, when Rains were moſt uſeful; it would have taken moſt place, and made them leaſt plentiful. For then the Earth it ſelf would have been hotteſt and drieſt, and the Subterraneous Abyſs moſt exhauſted.

Nor can theſe general Rains be pretended to come from the diſruption of the Abyſs; as if the fall of the Earth had cauſed ſuch extraordinary commotions in the Air, or convulſions of its Regions, as made them every where to pour down Waters. For * the *Theory* will have the Rains to be antecedent to the diſruption. *I do not ſuppoſe the Abyſs broken open till after the forty days rain.* But then this is moſt directly againſt Scripture again; for that plainly affirms the contrary; that the Fountains of the great Deep, and the Windows of Heaven were both opened upon *one* day, *Gen.* 7. 11. *In the ſix hundredth year of Noah's life, in the ſecond month, the ſeventeenth day of the month, the ſame day were all the fountains of the great deep broken up, and the windows of heaven were opened.* So that in the ſame year of *Noah*'s Life, and in the ſame Month of that Year, and on the ſame Day of that Month; the Fountains below, and the Windows above, were both ſet open; that the

* *Ibid.*

the Waters issuing out of both might raise the Deluge.

6. Let me add, in the next place, That it is a known Question, that has been moved by Writers of all sorts, Ancient and Modern, *Jewish* and *Christian*, Divines, Historians, Chronologers, *&c.* at what time of the year the Flood came in. *Josephus* (for instance) will have it to happen in *Autumn*; others in the *Spring*; and they give their reasons for it. The Question does manifestly proceed upon inadvertency; their not minding that when it was *Spring* in one part of the World, it was *Autumn* in another. And the like Question is put by Writers, and bandied among them, touching the Creation; at what time of the year that great Work was done. But somewhat more improperly, there being no Seasons of the year, before the Creation. Now this being the general Judgment of the Learned; *That the year had Tempestival Changes, from the beginning*, even the same that it has now (as these Questions import:) from hence it may be inferred, that they never dreamt of this Position of the Earth, or a Perpetual Æquinox; but were all of the contrary perswasion or common Opinion.

7. As for the *Authorities* that are made use of to establish the Doctrine we are upon; if they be examined, they will hardly be found to speak home in the case. For though in the *Contents* of the *Tenth Chapter* of the *Second Book* of the *Latin Theory*, it be thus declared; * *the last Article concerning the right Situation of the first Earth, is establisht by the sentences of Philosophers*: yet in thei

* *Articulus ultimus de situ recto telluris primæ —— Philosophorum Sententiis — stabilitur.*

their Sentences alledged in that Chapter be well consider'd; they will appear to be too weak and insufficient. I shall set them all down fully, to avoid suspicion of perverting or misrepresenting them.

The first is taken out of *Plutarch*, and delivered by him, as the joint Opinion of two ancient Philosophers. *(a)* Diogenes *and* Anaxagoras *think, that after the World was constituted, and living creatures were brought forth out of the Earth, the World in a manner was inclined towards its Southern part, of its own accord. And that this perchance was done by providence, that some parts of the World might be inhabited, and others not, by reason of cold, heat, and convenient temperature.* But this will do the *Theory* little service, it rather fights against it; For the *Inclination* here, is said to be made by Providence, that some of the Worlds parts might be οἰκητὰ κατ' εὐκρασίαν, *habitable by reason of a good temperature.* Which agrees not with the *Theory*; for that holds the World to have been of the best temperature, before the Earth was inclined; insomuch that it knew no Season but Spring. And what then could mend its habitableness? Yet in order to *that*, the Earth was inclined, as the Citation intimates. And when in the Judgment of these Philosophers, the inclination of the Earth was to conduce to, or improve its habitableness; and according to the Tenor of the *Theory*, it would rather have been an hindrance or disadvantage to the same: it is apparent that this Allegation does rather cross, than confirm the *Hypothesis*.

(a) Διογένης καὶ Ἀναξαγόρας, μετὰ τὸ συστῆναι τὸν κόσμον, καὶ τὰ ζῶα ἐκ τῆς γῆς ἐξαγαγεῖν, ἐγκλιθῆναι πως τὸν κόσμον ἐκ τοῦ αὐτομάτου εἰς τὸ μεσημβρινὸν αὐτοῦ μέρος, ἴσως ὑπὸ προνοίας, ἵνα ἃ μέν τινα ἀοίκητα γένηται, ἃ δὲ οἰκητὰ μέρη τοῦ κόσμου, κατὰ ψύξιν, καὶ ἐκπύρωσιν, καὶ εὐκρασίαν. *vid. Theor. Lat. pag.* 291.

In caſe it be argued, That this *Inclination* might promote or mend the habitableneſs of the Earth, as it quenched the flame in the Torrid Zone, and reduced its intolerable, to a gentle heat: neither thus can the Paſſage be drawn to favour the *Theory*. For (ſay the Philoſophers) by vertue of this *Inclination*, ſome parts of the Earth were to be rendred, ἀοίκητα, *uninhabitable*, and that κατ' ἐκπύρωσιν too, upon the account of vehement *heat*. Whereas this very Inclination, was of neceſſity to be a qualification, or corrective, or indeed a perfect extinction of all furious burning in the Torrid Zone; as * the *Theory* owns. So that the Authority cited, is ſo far from eſtabliſhing the *Theory*'s *Hypotheſis* of the Earth's *Inclination*; that it will not be eaſily reconciled to it.

* *Terrâ autem diſſolutâ, & exinde mutato ipſius ſitu facieq; una deſiere Zonæ Torridæ intolerabiles æſtus & ſiccitates.* Ibid. pag. 213, 214.

Nor can it excuſe the matter with this Pair of Philoſophers, to ſay that they were blinded here with the common Error, and ran, for company, with thoſe that believed there was a Torrid Zone, when there really was none. For allowing they were ſo ſagacious as to diſcover this Secret of the Earth's *Inclination*; we muſt alſo grant that by the ſame quick-ſightedneſs they would clearly have diſcerned, that the effect thereof could not have been, ἐκπύρωσις, a ſcorching, raging, inſufferable heat, about the middle of the Earth; but a certain mitigation or quenching of the ſame.

The ſecond *Sentence* is that of *Empedocles*, which occurs in the ſame Chapter of *Plutarch*: * *Empedocles* teacheth, *That the Air giving way to the force of the Sun, the North inclined, the Northern part*

* Ἐμπεδοκλῆς, τοῦ ἀέρος εἴξαντος τῇ τοῦ ἡλίου ὁρμῇ, ἐγκλιθῆναι τὰς ἄρκτους, καὶ τὰ μὲν βόρεια ὑψωθῆναι, τὰ δὲ νότια ταπεινωθῆναι καθ' καὶ τὸν ὅλον κόσμον. *Vid. Theor. Lat. pag.* 291, 292.

being

being elevated, and the Southern ones depressed, and this happened by that means to the whole World. Here is a mighty effect produced, without a cause assigned; at least here is *non causa, pro causa*: the assignation of a cause altogether incompetent and not to be understood. For why should the Air yield to the force of the Sun, more towards the South, than towards the North, when his force was equal upon both the Regions at once? For he moving at all times exactly in the midst betwixt them, his influence must be exactly alike upon each: and therefore that it should cause the depression of one more than of the other, is a thing in the dark and unintelligible. But say the Sun had had power to displace the Earth, and by sinking one Pole of it, through such a *cession* of the Air, to have raised the other: yet then that this cession should not be in the Air, nor consequently this dislocation of the Earth till the Flood happened; is not to be thought. And therefore *this* Sentence favours not the *Theory* neither; for *that* has positively determined the time of the Deluge to have been the juncture of the Earth's declension or * *dislocation*: Whereas if the Sun had been the cause thereof, by working a change in the Air conducive thereunto; it must have been accomplisht very long before.

* Pag. 186, 195.

The next *Sentence* is *Leucippus*'s, thus delivered by *Laertius*, (a) *That the Sun and Moon are subject to Eclipses long of the Earth's inclining to the South. And that the Northern Regions are always Snowy, Frosty*, and Icy. But by *Plutarch* thus, (b) *Leucippus* was of the mind, *That the Earth verges towards*

(a) Ἐκλείπειν δὲ ἥλιον καὶ σελήνην, κεκλίσθαι τὴν γῆν πρὸς μεσημβρίαν. Τὰ δὲ πρὸς ἄρκτον ἀεί τε νίφεσθαι καὶ κατάψυχρα εἶναι καὶ πήγνυσθαι. *Vid. Theor. Lat. pag.* 292.

(b) Λεύκιππος, παρεκπεσεῖν τὴν γῆν εἰς τὰ μεσημβρινὰ μέρη διὰ τὴν ἐν τοῖς μεσημβρινοῖς ἀραιότητα, ἅτε δὴ πεπηγότων τῶν βορείων διὰ τὸ κατεψῦχθαι τοῖς κρυμοῖς, τῶν δὲ ἀντιθέτων πεπυρωμένων. *Ibid.*

the Southern Regions, because of the thinness or openness of them; for while the Northern parts are frozen with cold, the opposite are hot. To take off this, we need but reflect on what has been said already; for how could the Southern Pole of the Earth dip into the Air, by reason the Air at that Pole was *hotter* and more rarified, than it was at the Northern Pole, when the Sun cut his way most evenly betwixt both the Poles? Or if it could have been so, yet then the Earth must have lost its regular Position, and the Equinox have been turned out of being, many hundreds of years before the Deluge came; which is utterly inconsistent with the *Theory*.

Democritus his Judgment also is brought in, in these words; *(a) The Southern part of the ambient Air being the weaker, the bulky Earth did therefore incline that way. For the Northern Regions being evenly, but the Southern unevenly tempered; thence it was, that accordingly it sagged down, where it abounded with fruits and increase.* Here is nothing new, save this, That the Earth abounding most with fruits towards the South, the weight of those helped to bear it downward; and so sway'd it out of its Æquinoctial Site; which in truth is but a vain and unphilosophic Phancy. For first, How could the Earth be more fruitful at one Pole than at the other, when the Soil was alike; and so, alike fertile; and both the Poles were equidistant from the Sun? Secondly, If the Earth had been most loaden with natural Increments, about its South Pole; yet how could these have overset or poized it down, by making it the heavier? For they all proceeding out of

(a) Δημόκριτος, διὰ τὸ ἀσθενέστερον εἶναι τὸ μεσημβρινὸν τοῦ περιέχοντος, αὐξομένην τὴν γῆν κατὰ τοῦτο ἐγκλιθῆναι. Τὰ γὰρ βόρεια ἄκρατα, τὰ δὲ μεσημβρινὰ κέκραται. Ὅθεν κατὰ τοῦτο βεβάρηται, ὅπου περισσή ἐστι τοῖς καρποῖς καὶ τῇ αὐξήσει. *Ibid.*

of the Bowels of the Earth; She must be as heavy before they grew up, as after. Thirdly, If the Earth could have been cast or settled towards the South, by those fruits we speak of; yet still here would be violence done to the *Theory*, by shutting its continual prædiluvian *Æquinox* quite out of doors. For the Earth being most fruitful at first, and consequently its Products about the Southern parts, most copious; That Pole, by their ponderous burthen, must have been overpowered in the *beginning*, and the Earth sunk into that inclining posture in which now it stands.

Having thus taken account of these Philosophers Opinions, before we go farther, let us make a short stop here; only so long as to remark these *Four Particulars*, already hinted.

First, That they of them who are most express for the *Inclination* of the Earth, do not deny this *Inclination* to have been from the *beginning*, or very soon after.

Secondly, That they do not only not deny this, but implicitly affirm it, by their assigning such causes of it. For though they be improper and such as never were; yet had they been, and could they have produced the effect at all, they would certainly have done it in the beginning of the World.

Thirdly, That none of these Philosophers, do make the least mention of a continual Æquinox antecedaneous to the Earth's *Inclination*. And in case it should be urged, that their very asserting the Earth to be inclined does suppose it was once in such a Position, as was attended with a fixed Æquinox. In way of answer it is observable,

Fourthly, That there is one Notion, which runs through most of their Assertions, and sufficiently proves, that they could never think the Earth held such

such a Position, as to be capable of a constant and settled Æquinox. For they intimate that the Southern Air was more *thin*, and *weak*, and *yielding*, than the Northern; as being more *temperate* or *warm* than that. But had they believed that the Earth kept a *Right Position* to the Sun; and so had both its Poles equidistant from him; they must withal have believed the Air about both, to have been of the like temperature and consistency.

All which put together, makes it evident, That the cited Testimonies are not *satis illustria*, *clear enough* to do the *Theory*'s business: and that the Article of the *Right Situation* of the Earth (the cause of the supposed *Æquinox*) is not at all established, *Philosophorum Sententiis*, *by the sayings of the* aforesaid *Philosophers*.

But therefore we have not done yet. *Anaxagoras* comes in with a second Attestation, and witnesseth, *That the Stars* † *were moved Tholiformly from the beginning, so as the Pole always appeared about the top of the Earth; but afterwards it declined.* So *Diogenes Laertius* delivers his mind. And this may seem to be somewhat a better evidence for the Earth's changing her Site. But in way of reply it might be noted,

† Τὰ δ' ἄστρα κατ' ἀρχὰς μὲν θολοειδῶς ἐνεχθῆναι, ὥστε κατὰ κορυφὴν τῆς γῆς ἀεὶ φαινόμενον τὸν πόλον, ὕστερον δὲ τὴν ἔγκλισιν λαβεῖν. Vid. Theor. Lat. pag. 293.

First, That *Ambrosius* the Monk (a good Philologer) who translated *Laertius* into *Latin*; instead of θολοειδῶς, reads θολερῶς, and so it signifies the Stars to have moved unevenly from the beginning; that is, as they do now. But let θολοειδῶς, be the true lection. Yet then,

Secondly, *Aldobrandinus* renders that, *turbulente*, *unsteddily*. And so makes the Philosopher speak the same sense that *Ambrosius* does, in a different word. But we will go farther still, and suppose,

Thirdly,

Thirdly, That *Anaxagoras* meant, that the Stars were carried about *instar Tholi, after the fashion of a* Cupulo (of which kind of Figure was the *Pantheon* at *Rome*, and therefore *Dio* calls that Temple θολοειδῆς, *Tholiform*) yet then might he not mean withal, that they imitated this Figure in their motion, only so far as the κορυφὴ, or Pole of the Earth, by being near to a direct Situation under the Pole of the *World* (not of the *Ecliptic*) would permit it to be done: For the *Declination* he here speaks of, we cannot understand so well with reference to the Pole of the Ecliptic; because he calls it the Pole *simply*, which denotes the Pole of the Æquator. And about this Pole indeed several Stars or Constellations, as the two *Bears*, the *Dragon*, *Cepheus*, *Cassiopæa*, &c. do move *Tholiformly* at all times: the Pole still appearing about the κορυφὴ, or *top* of the Earth; that is to say, about the Pole of it. And that these and other Stars thereabouts, *had declined* since the *beginning*, he might well be of Opinion; inasmuch as the *Vertex*, *top*, or Pole of the Earth, might in his time have suffered a considerable declination from the Pole of the World. According to which declination, the Fixt Stars seem to advance in Longitude. Insomuch that *Aries* hath passed into the *Dodecatemorium* or place of *Taurus*, in the *Zodiac*: and *Taurus* into that of *Gemini*, and so on. But then this is such a Declination, as does not at all imply, that the *axis* of the Earth was ever in a Parallelism with that of the Ecliptic; but only that it was once in a *nearer* Parallelism to the *axis* of the Æquator, than *Anaxagoras* found it in his days. And so the *Declination* he meant, might be quite different from that we contend about: which Astronomy imputes to the wallowing of the Earth, in its annual motion.

If

If this will not satisfie, I have one thing more to offer. Grant that *Anaxagoras* should mean that very Declination, which the *Theory* would have him: yet this truly would contribute little towards the Proof of the thing. For he was a Man as like to be heterodox; as like to broach and maintain false and groundless Opinions, as any of the learned Ancients. This perswasion concerning him, I build upon a wretched Foundation of his own laying: I mean that abominably gross, and shamefully absurd Assertion of his; That an huge Stone by the River *Ægos* in *Thracia*, fell down from the Sun. An extravagance so childish, and ridiculously unreasonable, as might justly give a wound, and a very mortal one, to his Philosophic reputation; and make the World conclude, that as to Skill in Astronomy he did not exceed. *Laertius* remembers this; and tells how *Euripides* his Scholar did hereupon call the Sun, *χρυσέαν βῶλον*, *golden Glebe*. *Plutarch* also mentions it in the Life of *Lysander*: and assures us that as the Stone was shown for a wonder; so it was venerable and of high esteem. * *Pliny* relates the matter more largely. But in case we should believe it, says he, (that a Stone could descend from the Sun) *farewell the knowledge of Natures Works, and welcome Confusion.* A very proper Reflection or Inference.

* *Nat. Hist.* *l.* 2. *c.* 58.

Nor is this to be lookt upon as a meer slip in *Anaxagoras*, or an unlucky error upon which he stumbled by chance. It must be his settled and approved Judgment: and I make it out thus; It is very agreeable to other Notions of his, or to the strain or *genius* of his Philosophy; Witness that strange way he invented, for *generating* the Stars.

Stars. For he thought *that the* * *ambient æther being of a fiery nature, did by its rapid circumgyration snatch up Stones from the Earth, and by burning them turn them into Stars.* According to the rate of which Philoſophy, that Stone of which the Sun was delivered, might poſſibly be a Star. And to this *Diogenes* very gravely ſubſcribes. For he roundly pronounces, τὸν ἐν Αἰγὸς ποταμοῖς πυροειδῶς κατενεχθέντα ἀστέρα πέτρινον: *That it was a ſtony Star which fell like fire at the River* Ægos. Which whoever can think, will not ſtick to credit † *Plutarch*'s Story. of a Lion, that in *Peloponneſus* fell down from the Moon: he being flung off thence ὑπὸ ῥύμης, *by* ſome *violent agitation* which ſhe ſuffered. Such was the Philoſophy of that Age.

* Τὸν περιεχόμενον αἰθέρα, πύρινον μὲν εἶναι κατὰ τὴν οὐσίαν, τῇ δ' εὐτονίᾳ τῆς περιδινήσεως ἀναρπάζοντα πέτρους ἐκ τῆς γῆς, καὶ καταφλέξαντα τούτους ἠστερικέναι. Ita refert *Plutarch. de Placit. Phil. l. 2. c. 13.*

† *De fac. in orb. Lun.*

This I have noted, not to diſparage *Anaxagoras* or *Diogenes*; but only to ſignifie, That where they ſtand alone, or are more poſitive than others in aſſerting any dark or doubtful Opinion; we have no reaſon preſently to run over to them, and to lay the ſtreſs of our belief upon their Authorities: eſpecially when in ſo doing, we muſt walk contrary to the whole World of the Learned at once. Yet ſo it happens, that the moſt likely evidence which the *Theory* brings in for the Earth's *Declination*, and ſo for its *Right Poſition*, and *Prædiluvian Æquinox*; is borrowed of theſe two Men.

8. Two Queries ſhall ſhut up this Chapter. The firſt is this, Suppoſing an *Æquinox* in the beginning of the World, *would it (in likelihood) have continued till the Flood?* For truly even a little matter might alter it; the condition of the primitive Earth conſidered. Indeed when it was newly formed, the upper Orb of it ſat as cloſe to the Abyſs as *that* did to the lower Orb. But

in process of time, the Waters of the Abyss were so exhausted, that betwixt the inside of this superiour Orb, and the surface of those Waters, there was an huge cavity; the Orb we speak of, hanging hollow round the Waters under it. But then it being so pendulous, and of so oblong a Figure as is here * represented; methinks

* See the same Figure in the *Latin Theory*, pag. 46.

Fig: 3 Pag 186

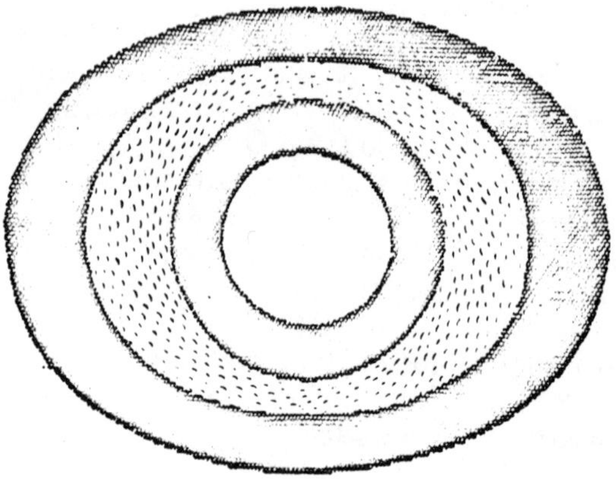

it might easily have been inclined towards the North or South (long before the Deluge) by some preponderancy or over-poize, that might have happened at either of its Poles, and determined it respectively. Thus (for

(for inſtance) put caſe that the Waters which fell in the frigid Regions had ſoaked into the Earth more at one Pole than at the other; or had flowed leſs freely (by reaſon of worſe Chanels) from one Pole than the other; or that upon ſome account they had been obſtructed or dammed up by the bordering Inhabitants, at one Pole more than at the other; and by their confined preſence there, had made it ſomewhat heavier than the other: it being in a juſt *Æquilibrium* before (and ſo eaſie to be ſway'd) muſt it not have ſunk immediately at that Pole, ſo overloaded; and receding from the level of its own Orbit, have ſtood in a leering or wry Poſition to the Sun? The Earth's own fruits was able thus to ſettle or draw it down, according to *Democritus*. And therefore it needs not ſeem ſtrange that ſuch an exceſs of Water ſhould do it; weigh it down from a true libration, into an oblique or inclining Poſition. Let us but remember what the great Mathematician, upon no ill grounds, once declared; δὸς πῦ ςῶ, &c. *give me but where I may ſet my foot, and I'll remove the Earth*; and what we have ſuppoſed will not look like an improbable thing.

The Second Query is this, Granting there was ſuch an *Æquinox* in the firſt World, *would not the natural day, towards the latter end of that World, have been* longer *than in the former periods of the ſame?* For while the outward ſhell or ſphere of Earth, was contiguous with the Abyſs; it ſeems very likely that it was carried about with more celerity, than it could be afterwards, when that contiguity ceaſed, by reaſon the Waters of the Abyſs were exhaled. And in caſe that external *Cortex*, the then habitable Earth, did abate of its diurnal Motion, upon loſing its contiguouſneſs with the

Abyss it inclosed; and the wider the distance grew betwixt them, the slower was its rotation; which must follow, if the failure of the contiguity we speak of did at first retard its gyration: then the days just before the Flood, must of necessity be longer than ever they were in the prediluvian World, supposing Day and Night be made by the Earth's turning upon its own *axis*. Especially if the Moon came late into the Earth's Neighbourhood. For then she being to be carried about in the exterior part of the Earth's *Vortex*, would have slacked its Motion; as an heavy Clogg hanged upon the Rim of a Wheel, makes it turn more slowly. Yet that the days just before the Flood were of no unusual length, is evident in the very Story of the Flood; the duration of which we find computed, by Months consisting of *Thirty* Days apiece. Whereas had Days been grown *longer*, fewer of them would have made a Month.

CHAP.

CHAP. IX.

1. *The* Oval Figure *of the Primitive Earth excepted against, from the nature of that* Mass *upon which it was founded.* 2. *And from its* Position *in its Annual Motion.* 3. *As also from the* Roundness *of the* Present *Earth.* 4. *Which Roundness could not accrue to the Earth from its* Disruption, *in regard that would have rendred it* more Oval *still, in case it had been Oval from the beginning.* 5. *Or at least would not have made it* less Oval *than it was.*

1. AMong the several Properties of the Predeluvian Earth, there was none more needful than its *Oviformity*. But as needful as it was, it seems a thing improbable. The necessity of it, is apparent from its Usefulness: and that was as great as can well be imagined: For it was to be as an *Aqueduct* to the first World; or a general Instrument of deriving Waters, into all the inhabited Quarters of it. So that without it, according to the Laws of this *New Hypothesis*, the Earth would have been outwardly but a lump of Sand, and as miserably barren as any piece of Wilderness the worst *Arabia* has. And yet if we attend to the first Earth's Origination, how could it be of an *Oval* Shape?

For a liquid Mass, having its Center in it self, and being of a Substance equally yielding in all its parts, and likewise equally compressed by an ambient body: must of necessity be equally extended in all the lines of its circumference; that is, it must be exactly

exactly round or spherical. For why any piece thereof should thrust up higher, or shoot out farther from the common Center, than the rest; there can be no reason given; Unless, according to the *Hylozoic* Philosophy, we should suppose there is Αὐτοκίνησις, a principle of Life or *self-movency* in matter; which indeed is to exalt it above its capacity, and to give it a property that destroys its nature.

And were not these the very circumstances of that Mass, whereon the Primitive Earth was founded? For,

First, It was *self-centred*; and by vertue of its proper Centre, so entirely coherent and united, that no parts of it had the least tendency towards jetting out, or flying off from the whole: but by the Laws of gravity, were all impregnated with the contrary determination, a nitency inward or downward towards the Central Point. And then,

Secondly, It was *Liquid* also; and so of a yielding temper or consistency. Ready to give way to the lightest pressures, and by a forward pliantness, to fall into that Figure, into which the circumfluent Air would fashion it. For that Element alone (or a thinner than that) but by moving and gently gliding upon it, might easily smooth it into perfect evenness; provided it did but encompass it around, and so was capable of slicking it by a general levigation. And therefore,

Thirdly, It swam in such a fluid Element, as did *so* environ it, grasping it on all sides with a soft compression. So that during its fluitation in that surrounding and gently constringent *Medium*, it could not but be of a truly *Globular* Form. Which admitted; the Primitive Earth must needs be be ſo too, and not *Oval*; as being caſt upon this *Globous* Mould.

But to this it is oppoſed, That the Liquid Maſs whereon the firſt Earth was built, was not *quieſcent*. So it *might*, yea, it *muſt* have been truly *ſpherical*: And the *Theory* it ſelf owns as much. (*a*) *I nothing doubt*———*but a maſs of Water will naturally make it ſelf into a ſpherical Figure, about its own centre; if ſo be it reſts immovable and quiet.* But then it adds; (*b*) *But in caſe it be turned ſwiftly about its Center, by that agitation it will neceſſarily make it ſelf oblong, and become of a Figure ſomewhat Oval; juſt as when Waters are puſht forward in a Veſſel; or in ſome part of a Sea or Lake, are driven by a Wind toward the Shores; we ſee the Waves ſtretch themſelves out long-ways.*

(*a*) *Nihil dubito* —— *maſſam aquæ ſe naturaliter conformaturam in figuram ſphæricam circa ſuum centrum; ſcilicet ſi illa maſſa ſit globus aqueus immobilis & quietus hæreat.* Pag. 193.

(*b*) *Si vero volvatur rapidè circa ſuum centrum, ſe oblongabit neceſſariò ex agitatione illa, & defluet in figuram præter-propter ovalem; ut cùm aquæ in vaſe propelluntur, aut in aliquâ plagâ maris aut lacûs vento agitantur verſùs littora, fluctus in longum ſe extendere videmus.* Ibid.

In anſwer to which, let it be confeſſed, That the Liquid Maſs, on which the Earth was raiſed, was rolled about; and that very ſwiftly upon its own Centre. Yet that by vertue of its gyration it ſhould be ſhaped into an Oval faſhion; was not at all neceſſary; nor will the Inſtances brought in, prove it was ſo, there being no parity or juſt proportion betwixt the ſeveral Caſes. For, for Waters to be forced an end by the external violence of Winds (where the impreſſion propelling them is *ſuperficial*, and their motion *progreſſive*;) is a different thing from their circumrotation in one entire *Moles*; where they turn only with a natural and moſt even Courſe, carrying the ambient body (whereby they are circumſcribed and helpt to keep their

their Figure) round along with them. For thus we see, that notwithstanding the Earth turns so swiftly, that every point in its circumference under the Æquator, moves at the rate of fifteen Degrees (nine hundred Miles) an hour; yet the finest Sand upon the surface of the Earth, or the lightest Dust upon the tops of the Mountains, is never dissipated or disturbed in the least, by this whisking circumvolution. Whence we may gather (the case being much the same) that the whirling Globe of Water, was so far from a necessity of growing oblong, by its rotation; that that very thing might contribute to preserving it in a Globular Form.

But therefore let us hear what the *Theory* says further, and more distinctly yet, touching the *Cause* of the *Oval Figure* of that Mass of Water, which was the basis of the primitive Earth. It speaks it fully in these words, (*a*) *Nor is the reason of this Figure obscure, in a Globe of Water which is moved circularly; for the Mass of Water being much more agitated under the Æquator, than the Waters towards the Poles, where it passed through lesser circles; those parts which were most moved endeavouring to recede from the Centre of their motion, when they could not quite spring up and fly away, because of the Air which lay upon them on every side, nor yet could fall back again as being checked and resisted by that Air: they were unable to free themselves any other way than by flowing down to the sides: for Waters being pent, do flow that way where they find easiest passage; and from*

(*a*) *Neque ratio hujus figuræ in Globo aqueo circulariter moto, obscura est: cum enim moles aquæ sub Æquatore multo magis agitaretur, quàm aqua versus polos, ubi minores circulos peragerent; partes illæ maximè agitatæ, à centro sui motus recedere conantes, cùm sursum exilire & avolare non potuerint propter incumbentem undique aërem, neque multum refluere sive ejusdem aëris renixu & resistentia, non aliter se liberare valerent, quàm defluendo ad latera: Aquæ enim impeditæ quocunque reperiunt aditum & faciliorem motum, eo fluunt; & ex illo defluxu aquarum ad latera, & exoneratione partium mediarum circa Æquatorem, globus aqueus devenerit aliquantulum oblongus.* Ibid.

that

that flowing down of the Waters to the ſides, and diſburthening of the middle parts about the Æquator, the Globe of Water might become ſomewhat oblong. So that the Cauſe of the Oval Figure, in the *Chaotic* Waters, ſeems (in ſhort) to be this. Their diſcharging themſelves, *defluendo ad latera, by flowing down to the ſides* or Poles of the Globe; upon their ſwelling or riſing up (by means of their rapid circular motion) about the Æquator. But granting the Waters did ſwell and riſe thereabouts (which yet would admit of diſpute;) againſt this piece of the *Theory*'s *Hypotheſis*, it may be thus excepted; Either the Waters did flow down to the ſides or Poles of the Globe, till it became Oval; or they did not. If they did not flow down ſo long, the *Hypotheſis* fails, and the watry Maſs could never be Oval. If they did flow down ſo long, then they muſt flow down, till they flowed *down, upwards.* (Pardon the abſurdity of the Expreſſion, the abſurdity of the thing occaſions it.) For the Polar parts of the watry Maſs, as it became Oval, were the higheſt, being moſt diſtant from the Centre. And yet from the Æquator they did *defluere ad latera, flow* **down** *to the Sides* or Poles. Which that they might do, it was abſolutely neceſſary that the parts about the Æquator ſhould be *higheſt*: elſe the Waters in flowing to the Poles, would have been ſo far from flowing *down*, that contrary to their natures, they muſt have riſen up above their Source. And yet as abſolutely neceſſary again it was, that the Polar parts ſhould be *higheſt* at laſt; otherwiſe the watry Maſs could never have been made of an Oval Figure. And yet if it were made into that Figure, by the Waters flowing *down* (as the *Theory* ſays) from about the

Cc Æquator,

Æquator, to the ſides, or Polar parts; then a *third* thing will be as neceſſary as either of the *two* mentioned; namely, That the Waters (as was ſaid before) ſhould flow *down, upwards.* So that it is as unlikely, that the Maſs of Waters was ever of an *Oval Form*; as it is unlikely that a Contradiction ſhould be true; or that the Element of Water ſhould, *of it ſelf*, perform a motion, which is beyond its power, by being above or againſt its nature. I ſay, *of it ſelf*; for however there might be violence (that of the circular motion) in making it to ſwell about the Æquator; yet when once it was riſen there, it was left to it ſelf, as I may ſay; all farther force was taken off it, and it might follow the duct of its own Principles of *Gravity* and *Fluidity.* And accordingly it is ſaid by the *Theory*, *ſe liberare, to free it ſelf* (from that force which it ſuffered in receding from its Center, or riſing up under the Æquator) *defluendo, by falling off* or *flowing down*; a proper expreſſion of the true *natural* motion of Water. But then if the place it fell or flowed *to*, was higher than that it fell or flowed *from* (as in this caſe it muſt prove, before the watry Globe, by *Defluxion* of its Water, could be made *Oval*) it is evident that the Water by a natural motion, or *of it ſelf*, did perform a courſe againſt its nature. For when it *flowed down* (or is ſaid to do ſo) and according to its own nature ought to have done ſo; in reality, and according to the reaſon of the thing, it *flowed up.* Nor indeed could it poſſibly do otherwiſe, to produce the great effect pretended; unleſs it were poſſible for an Oval body to be higheſt in its *middle* parts. And then truly (but upon no other terms) the watry Globe *might become oblong, ex illo defluxu*

aquarum

aquarum ad latera, & exoneratione partium mediarum; from that flowing down of the Waters to the sides (which the *Theory* mentions) *and the disburthening of the middle parts* of it.

Now if the watry Mass, upon which the ingenious *Theorist* founds the first Earth; could not be made *Oval*, in the way he has invented: then neither could that Earth be of an *Oval Figure*, it being bound to put on the same shape which the Water had.

2. Very improbable it is also that the first Earth should be Oval, considering its Position or Direction in its Annual Motion. For that was such as could not well consist with its Oval Figure. In it, its Poles are said to have pointed always to the Poles of the Ecliptic: and so it would have been directed not unlike to Ships swimming side-ways. Now put a Ship, which is an Oval body, into the smoothest stream imaginable, and lay it cross that stream; and see how long it will keep in that Position. Will it always hold it? No, nor for any considerable while: but by degrees will quickly wind and fall in to swim long-ways with it; and continue mostly in that posture, as suiting best with its own shape and the course of the Waters. And truly that an Oviform Earth should lie cross-ways in the æthereal Chanel, and be carried round the Sun for Sixteen hundred years together; and not change its site in compliance with the tendency or stream of it; seems very strange, if not impossible. Especially when that Earth was thin comparatively, and hollow like a Shell; and so more light, and ready to verge or be drawn aside from its supposed primitive situation.

The Present Earth, though generally allowed to be of a spherical Figure; and also of a solid composure throughout (unless at its Centre ;) and likewise (according to the *French* Philosophy) to be held by a particular hand of Nature, in its inclining posture (which must be more easie to be kept by a *round Earth*, in the *Medium* which carries it; than a *Right Position*, by an *oblong* one:) is yet subject to wallowings in its Annual Motion. And how then can it be thought, that the First Earth, which was oblong, and had not that hand to hold it steady; could preserve its *axis* in a constant parallelism to the *axis* of the Ecliptic, till the time of the Flood? It would rather have turned end-ways in the Celestial Stream, and have stood for the most part in that direction; as best agreeing with its own Form, and the vehicular Current wherein it floated. And so (its axis, by force of the æthereal matter, being wrought into a coincidence with the Plain of the Ecliptic; and the Ecliptic like a Colure, passing through its Poles) while its Poles would have lookt East and West, and its Diurnal revolutions have gone North and South: it would have brought such a confusion into the Heavens and Earth at once, as is not easie to be expressed.

3. And that the First Earth was not Oval, methinks may, in some measure, be gathered from the Roundness or Sphericalness of the Present Earth. For this Terraqueous Body on which we dwell, is of a Spherical Fashion. So *Anaximander* thought, and also *Pythagoras*, *Parmenides*, and others of old, as well as all of later days. And as much is fairly inferrible from several things. As,

First,

First, From its *Conical* Shadow. Which Figure, * *Zeno* (almost Two thousand Years since) noted the Shadow of the Earth to be of. And a common Argument for the Proof of it, is fetcht from the Moon. For in whatever place she has at any time entered into an Eclipse, or emerged out of the same; and whatever part of the Earth, during any of her Eclipses, has been turned to her, still it has been observed, that the Shadow cast by the Earth upon her *Discus*, was always Circular; which argues the Earth it self to be Globular. And that it is so, may be inferred,

* Τὴν γῆν ϗ κωνοειδῆ σκιὰν ἀποτελεῖν. Diog. Laert. *in vit.*

Secondly, From the Place of the Waters. For were it Oval, they would not fail to retire out of the Seas near the Poles; and running down towards the Æquator of the Earth (which would be the lowest part of it) settle themselves around it, in the middle Regions thereof. But instead of this, we see the Waters are so far from drawing off from the Northern Seas about the Pole, that they abound most, and are deepest there; nor do we know of any thing but vast and deep Waters about the South-Pole neither. Whereas, I say, were the Earth *Oval*, and so the Poles of it highest, the Waters must necessarily have settled about the midst of the Earth, there being the lowest place, and so the properest for their Situation. And so the Sea in Figure would have resembled an Hoop; or as a liquid Zone would have encompassed the Earth, and divided it into two Hemispheres, in the same manner that some worthy Ancients conceived it did; for want of better Skill in Geography.

Thirdly,

Thirdly, If the Earth were Oval, Navigation towards the Poles, beyond such a Latitude as bounds the Sphericity of the Earth, would be extreamly difficult, if not impossible. For then in such a course, Ships must steer up hill, and climb, as it were, all the way they swim, as sailing in a perfect ascent. But where would be Winds strong enough to heave them up such watry steepness? Or in case they had sufficient strength to do it; yet would not the Vessels rather pitch into, and run under the Waters that bear against them than drive up upon their rising surface? And let but the blustring Gales which push them upward, cease; and would they not forthwith stop? Yea, immediately tack about, and (being left to themselves) settle down towards the Æquator again. But we hearing of no such difficult sailing up the Polar Seas; nor such retiring of Ships down to the Æquinoctial ones; have still more reason to believe, that the Earth and Water make a True Globe. And grant that these Arguments will not perfectly *demonstrate* the Earth to be Spherical; yet they being of more force to prove it is so, than any ever brought to prove it otherwise; we have reason to acquiesce in the received Opinion.

4. But to this it may be answered: In case the Earth be Round or Spherical *now*, this is no good evidence that it was so at *first*. It might *then* be Oval or *oblong*, and its present Roundness may be owing to its Disruption. I reply; Admit the Earth was oblong before the Disruption, and the falling in of its outward Orb, could hardly reduce it into a Spherical Form; but would rather have made it *more oblong* still. For the Orb we speak of must (in

(in likelihood) break and fall in firſt, about the Æquator, or Middle parts of it. For there it was moſt heated; and there it was moſt cracked; and there it was moſt hollow underneath; the Waters of the Abyſs being much exhaled. And theſe parts falling, thoſe whereabouts the *Tropics* are now, might fall ſoon after them. Whereas the Poles of this Orb, being turned with a ſhorter or narrower Arch, were much the ſtronger. And then being remote from the Sun, and continually wet, were nor diſpoſed to break at all, through drineſs and brittleneſs, as the Regions about the Æquator were. So that the Poles might remain whole, and keep thoſe very places almoſt which they held before. For as for their ſinking lower, and coming much nearer together, than they were, it was not likely; becauſe that huge Circle of Ground, which fell in about the Æquator and Tropics, would have intercepted and hindred them. For though the Poles were hollow, they could not ſlip over the Earth which fell in betwixt them, and claſp it in their cavities; in regard they were not wide enough. For the Orb being Oval, was narroweſt towards the Poles. So that the falling in of the Earth, muſt have rendred it rather *more*, than leſs Oval; While the Poles of it would have continued at their uſual diſtance almoſt, and the intermediate Regions, by dropping into the Abyſs, would have been contracted into ſtreighter Dimenſions of Circumference.

5. Or ſay the Diſruption of the Earth, would not have made it *more* Oval than it was; yet ſurely it would not have made it *leſs*. For as the Earth, in all probability, would have broke in firſt about the Æquator (for the reaſons alledged) ſo thoſe Fragments

Fragments being nearer the inward Earth, than the Polar parts; would sooner have reached it in their fall, than these could have done. Especially considering these Polar parts (according to what was said before) must have fallen entire in two vast Caps as it were. For so they would have contained such abundance of Air, as must have rendred their descent very slow; much slower than that of the Æquinoctial, and Tropical Fragments. Which being of quite another fashion, that did not inclose the Air so much, would have descended a great deal faster. Insomuch that before the Polar Hemispheres (let me call them) could have got down to the interiour Earth; all the ground that fell in about the Æquator and Tropics, would have been settled there, and fit to receive those mighty Hemispheres, when they should have come and whelmed themselves *whole* upon it. Or grant they should have *broke*, by pitching upon that vast heap of Earth, which fell down betwixt them; yet there they must have laid in a confused posture where they flew in pieces, and so would have helped to make the Earth oblong. In a word, suppose they did sink down as far proportionably towards the common Centre, as the Æquinoctial and Tropical parts did; yet if they sank no farther (as indeed why should they, all circumstances considered?) the Earth in case it were Oval at first, must of necessity continue so.

CHAP. X.

1. *That there were* Mountains *before the Flood, proved in way of* Exception *to the* Theory, *out of* Scripture. 2. *And that they could not be made by the* Falling in *of the first Earth, argued from the* Mountains in the Moon. 3. *And from the* Opinion *of the* Talmudists, *and others.* 4. *How* Mountains *might arise in the very* beginning. 5. *There must be* Mountains *in the* first World, *because there were* Metals *in it.*

1. TWO Properties of the Prediluvian Earth we have done with; its *Continual Æquinox*, and its *Oval Figure.* We must now proceed to its next Property, or rather to the former Branch of it: *The exterior face of it was smooth and uniform, without Mountains.* But neither can this be asserted without some violence to the Inspired Writings. *LORD, thou hast been our refuge from one Generation to another. Before the Mountains were brought forth, and the Earth and the World were made, Thou art GOD from everlasting.* So we read, *Psal.* 90. 1, 2. Where the scope of the *Psalmist* being to set forth GOD's Eternity, and his early Providence over his People; he declares of him, That as he was always a Shelter and Protection to them from Age to Age; so he existed before the Creation: even *before the Mountains were brought forth, and the Earth and the World were made.* Where his ranking the Production of the Mountains, with the Formation of the Earth and the World; speaks them *coæval* with the same.

And (which is not unworthy of remark) *Moſes* (ſays the Title of it) compoſed this *Pſalm*; to whom the Riſe of all things, and the Order of their riſing into being, was better known, than to any Man born. Yet this *Moſes*, as he illuſtrates GOD's Eternity *à parte ante*, by his Preexiſtence to the Univerſe; ſo he meaſures his moſt timely care over his Church, as much by the *Mountains* duration, as by the duration of the *Earth* or *World*. Thereby giving us to underſtand, That the one is as good a Rule as the other; as bearing the ſame date of Exiſtence, and iſſuing forth into being, not by a far diſtant Succeſſion, but all together, as faſt as nature could permit. And however *ſome* Mountains might be produced long after others; yet that will make nothing againſt us, if we do but ſuppoſe the *Pſalmiſt* to ſpeak of the *Earlieſt*. This, by the way, does ſufficiently confute the *Peripatetic* error touching the Worlds *Eternity*. For if GOD was, בטרם ארץ ותבל. *before the Earth and habitable World*; 'tis certain the World had a beginning, and could not be *from everlaſting*, as he was.

And the ſame *Moſes* makes mention not only of *laſting* Hills, but, of *ancient* Mountains, *Deut*. 33. 15. But had there been no Mountains till the Flood, he would ſcarce have given them that Epithet, as being but few ages older than himſelf. I confeſs, קדם. ſignifies, *the Eaſt*, as well as *ancient*. And the *Samaritan* and *Syriac*, *Pagnin*, and *Montanus*, render, הררי קדם, *the Mountains of the Eaſt*. But not ſo well. For *Ephraim* and half the Tribe of *Manaſſeh*, *Joſeph*'s Poſterity (mentioned *ver*. 17.) in whoſe Land the Bleſſings of theſe Mountains are here propheſied of by *Moſes*, were planted in that

that Division of *Palestine* called *Samaria*: which being on this side of *Jordan*, and upon the *Mediterranean* Sea, lay towards the *West*; and consequently its Mountains could not be called *Mountains of the* East. And as for the other Half of *Manasseh*, though they were seated beyond *Jordan*; yet I do not find that the Mountains in their allotment, were known by the name of *Mountains in the East*. Or if *Basan*, and some of the Hills of *Gilead*, belonging to this half Tribe, might be so denominated: yet the whole Tribe of *Ephraim*, and the other half of *Manasseh*, had nothing to do with them; and so *they* could not *so well* be the Mountains here pointed at in the Prediction. Inasmuch as they were possessed but by the smaller part of *Joseph*'s Off-spring; whereas in reason the Prophecy should respect the *Major* part of his Seed, and so refer to those Mountains on the West side of *Palestine*, where the whole Tribe of *Ephraim*, and the other half of *Manasseh* were settled. And therefore when the *Arabic*, *Vulgar*, *Junius* and *Tremelius*, and others, render, קדם. *ancient*; they give it its most proper Signification in this place, and such as makes the best or truest sense. For the Mountains of *Gilead*, and *Basan*, as well as *Mount Ephraim* (a ridge of Hills crossing the Country of that Tribe; several parts whereof were *Gerizzim*, and *Ebal*, and the Hills *Tsuphim*; the Hill of *Phineas*, of *Gaash*, of *Salmon*, and of *Samron*, whereon *Samaria* stood) were all of them, though not *Eastern*, yet *ancient* Mountains; such as might take their beginning with the Earth it self, or immediately after. And therefore the *Septuagint* calls the Mountains here, Ὄρη ἀρχῆς, *Mountains of the beginning*. And that most fitly; for, קדם. signifies, *beginning*, as well as *antiquity*. And accordingly we read, מִקַּדְמֵי־אָרֶץ, *from* or *before the beginnings of the Earth*, *Prov.* 8. 23. And so the *ancient Mountains* in the Prophecy, are such as were extant *à primordiis seculi*, from the beginning of the World, in contradistinction to such as were casually raised, or artificially made.

Solomon also, in the same Chapter, speaks a remarkable word to our purpose. For declaring the Antiquity, or Eternity of THE DIVINE WISDOM; he there sets it out, by its existing *before GOD's Works of old*, *ver.* 22. and then particularly, *before the Mountains were settled*, *ver.*25. A most clear evidence, that the settling of the Mountains, was one of the earliest Works that ever GOD did. Else it could not have been sutable to sort it with those that are there recounted, nor would it have been proper to shew forth the Antiquity of WISDOM by it; and to argue that it was in GOD's Possession, *in the beginning of his ways*, because it was so, *before the Mountains were settled*. For if the Mountains were settled long after those other *works of old*, which *Solomon* specifies, as long after, as the general Flood was after the Creation; why should he place *this* work amongst them, and rank it with them, as one effected at the same time? And if they were settled, at the time of the Deluge, how could WISDOM's existing before they were settled, be a Proof or Illustration of its being possessed by the LORD *in the beginning of his ways*?

So that if we can but think, that Scripture is to be understood like other writings; that is, according to the common signification of its words, and the manifest drift or scope of its sense:. And if if we can but believe, that when GOD speaks with design to instruct us; He does it with the same freeness and sincerity, as a rational and honest Man, as a kind and open hearted friend would do, that means to discover his mind unto us; we need not desire more pregnant Proofs of the Mountains just coævity with the Earth.

2. The same Scriptures that prove Mountains coæval with the Earth, are clear Evidences that they could not arise from the Disruption of the same. Which Opinion being so fairly encountred and overthrown by *Divine* Authorities; to pursue it farther may seem unnecessary. But yet that *Nature* may set its hand to the confutation of it, as well as *Scripture*; I will here put down One single Argument which the Moon affords us.

That she hath her Mountains as well as the Earth, is very evident. And also that they are *higher* than the Earth's Mountains: I mean not only comparatively, in proportion to her Bigness; but they are so, simply and absolutely in themselves, if we dare credit *Galileo.* Yea, they are not only higher than the Mountains of the Earth, but better than *four times* as high, as he undertakes to * demonstrate. Whence it must follow, either that the Moon was not formed and dissolved, the same way that the Earth was †; both which the *Theorist* owns her to have been; or else that her Abyss was deeper, and her outward Orb thicker by far, than was the Earth's (to make such prodigiously lofty Hills) and so that she was very much *larger* than the Earth: the contrary to which is most true and manifest. Or, lastly, (which is the case) that her Mountains were not the effects of her Dissolution (which

* *Astronom. Nun. pag.* 24, 25, 25.

† *Vid. Lat. Theor. pag.* 58. and also the *English* one, p. 110, 168, 170.

(which she never suffered;) but her *Native Features*, and such as she has worn ever since her Creation. But then why should it not be so with the Earth likewise? Or how can it be otherwise? For were it granted, That the Mountains of both did at first arise (as the *Theory* would have them) from the falling in of their respective exterior Orbs; it would be hard to assign reasons, why Mountains in the Moon, should be *four times* higher than any on the Earth; when the Globe of the Earth, is above *forty times* bigger than that of the Moon.

3. And that Mountains were in being before the Flood, and so could not result from the *falling in* of the Earth; we may learn, in some measure, from the *Talmudists*, even while they teach what is phanciful and extravagant. For they report, That many Giants saved themselves from the Flood upon *Mount Sion*. And *Josephus* intimates such another Tradition out of *Nicolaus Damascenus*: (a) *There is above* Minyada, *a great Mountain in* Armenia, *called* Baris, εἰς ὃ πολλοὺς συμφυγόντας, &c. *to which many flying in the time of the Flood, are said to have escaped.*

(a) *Antiq. lib.* 1. *cap.* 4.

As for the *First* of these Reports, it is wholly fabulous; nor can it be otherwise as being repugnant to Scripture and Reason. The *other*, though certainly false in the gross, may yet have somewhat of truth in it, as being a broken account of the Preservation of *Noah* and his Family, and the Story of their Deliverance mangled and disguised. For it being commonly believed, that the Ark rested upon the Mountains of *Armenia*; and that the Old World being Drowned, the New one was Peopled

Peopled by Men γένος δευτέρον (as *Lucian*'s word is) *of a second stock*, that came down from thence: this might give occasion to that formal fiction of a Multitude flying to *Baris*, and of their being saved there.

Now though one of these Traditions be absolutely false, and the other a Truth perverted and misrepresented; yet such things being talkt of in times of old, and at last put in writing, they do fairly witness what the Thoughts of Men were in former Ages, as to this matter; and that it was a current perswasion among them (who lived much nearer to the first World) that there were Mountains before the Deluge.

And such another piece of confused Forgery (out of sacred Story corrupted) occurs in (a) *Clemens Alexandrinus.* He took it out of *Plato*, and it speaks of a Flood to come. *But then again when the GODS drown the Earth, purging it with Waters; the Herdmen and Shepherds shall be saved on the Mountains; while they that are with us in Cities, are carried by Torrents into the Sea.* And that this was a Fragment or lame kind of Excerption out of the Holy Oracles; the Father himself signifies. For he presently indites the *Greek* Philosophers of Pilfering, and draws up this smart Charge against them; That they were a Pack of *ingrateful Thieves*, *who filched*, τὰ κυριώτατα τῶν δογμάτων, *the* chief *of their Opinions from* Moses *and the Prophets.* And if this Flood had not been greatly mistaken as to time, and so the Story of it set with its face the wrong way, it would have looked directly upon what we are asserting.

(a) Οτ' ἂν δ' αὖ Θεοὶ γῆν ὕδασι καθαίροντες κατακλύζωσιν, οἱ μὲν, ἐν τοῖς ὄρεσι διασώζονται, βουκόλοι καὶ νομεῖς. Οἱ δ' ἐν ταῖς παρ' ἡμῖν πόλεσιν, εἰς τὴν θάλασσαν ὑπὸ τῶν ποταμῶν φέρονται. *Strom. lib.* 5.

asserting, and given countenance to it. For then the Flood here mentioned, must have been that of *Noah*; and the Mountains of refuge for the Herdmen and Shepherds, must have been extant in the First World.

It is well known also, that many of the Learned Ancients have taught, that Paradise was situate upon high Mountains. And according to that Doctrine, there must be Mountains at the very first. And however some eminent Writers are of Opinion, That the Mountains were neither so *many* nor *great* before, as since the Deluge; yet none, I think, ever excluded them wholly, till then. And I durst appeal to the *Theorist* himself, if ever he met with any, that held the Earth was without an *open Sea*. Yet as many as suppose such a Sea in Nature, suppose Mountains too; and 'tis necessary they should as himself confesseth. The Consequence of which will be, That no Authority is to be brought, or heard, against the being of Mountains before the Flood; but such as is express against open Seas. And then I presume we may search long enough, before we find one.

I will only add that Traditional Story which is told of *Adam*; namely, how that after his Fall, and when he repented of his sin, he bewailed it for several hundreds of Years, upon the Mountains of *India*. Another plain intimation that there were Mountains in the beginning of the World.

4. Nor is it hard to conceive how they should be made then, as well in-land ones, as others. And that in such a way (to humour Philosophy) as Nature might have a considerable stroke in the Work. For though it be not for us exactly to understand the

the manner of GOD's Proceedings in this Case (whose ways in forming the Mountains, as well as in other things, *are past finding out*; and for Men to offer at a clear and certain Explication of their Rise, would be arrogant presumption; as if the Nut-shell of their Phancy, could contain the Ocean of the Divine Methods) yet with humblest Adoration of the ALMIGHTY's Infinite Power and Wisdom, and acknowledgment that he *could*, and 'tis like, *did* produce them another way; I will venture to guess he might do it thus. But I only hint, what it would require a large Discourse, to make out and confirm in every Particular.

The Earth, when it was first created, lay under Water (as the infallible Word informs us) till the Third Day; and on that Day, the Waters were gathered into one place. The *Alveus*, that is, or *Hollow* of the Sea being prepared, by GOD's pressing down the Ground (suppose) lower there, than it was in other places; the Waters fell violently into that Cavity. And as they were carried thither in a Natural Course, while by the force of their Weight they rolled downward: so they were help'd by a Power supernatural; 'tis like by the Influence of that Blessed SPIRIT, who moved upon them when they were first brought forth. Otherwise perhaps they could not have been so drained off the Earth in one Day, as that the dry Land should have appeared. Now the Earth, by this Collection of the Waters into one place, being freed from the load and pressure of them, and laid open to the Sun; the Moisture within it, by the heat of his Beams, might quickly be turned into Vapours. And these Vapours being still increased by the continued rarefying warmth from above;

at length they wanted ſpace wherein to expand or dilate themſelves. And at laſt not enduring the confinements they felt, by degrees heaved up the Earth above ; ſomewhat after the manner that Leaven does Dough , when it is laid by a Fire; but much more forcibly and unevenly. And lifting it up thus in numberleſs places, and in ſeveral Quantities, and into various Figures ; Mountains were made of all ſhapes and ſizes.

Thus we may conceive the *In-land* ones were produced ; which in ſome Countries were more, and in ſome fewer ; in ſome bigger, and in ſome leſſer ; in ſome higher , and in ſome lower ; in ſome again earlier , and in ſome later : according as the Nature of the Soil, the Vapours under it, and the Sun above it, contributed and concurred to the raiſing of them. And how a Ridge or Chain of Hills might be blown up at once, as well as one ſingle one ; how Mountains ſhould be hollow at the Roots, and in their higher parts, and full of Caverns ; how in time they might be dried, hardened , and turned into Stone in a great meaſure ; how ſome of them, through their weight and hollowneſs, might break and fall ; and in their hideous Fragments and diſorderly Poſtures, repreſent the ruines of an Earth ſunk into an Abyſs : and others might be eaten and worn away by Time and Weather, eſpecially by that Weather in time of the Flood ; and ſo become rough and craggy, and ſurprizingly horrid and frightful things will be obvious, or at leaſt intelligible, to thinking and Philoſophic Minds.

And that Mountains might be brought forth thus at firſt, or raiſed in the way we ſpeak of, will ſeem more likely ſtill, in caſe we conſider how Hills oftentimes

times have been thrown up by Earthquakes: where though the Causes were not the same, they were very like them, or analogous to them. The Earth also at first, was most disposed or liable to these Effects; I mean, to have Mountains made out of it. For then the Soil (being destin'd and prepar'd to be the common *Seminium*, *Seed-plat* and *Nursery* of all sorts of Vegetables, and of some living Creatures) was soft, and light, and unctuous; and so of a very yielding Nature. The Pores of it also were then close shut up, as having never been opened by Sun or Winds. By which means the Vapours imprisoned in it having no manner of vent; when they became strong enough (by their daily increase) might easily cast up huge quantities of Ground thereby to free themselves, and get loose from under them.

Nor need we wonder, that sometimes a Valley betwixt two Hills, should be lower than the common surface of the Earth. For the matter of those Hills being spewed up from under that Tract between them; the ground must there sink down in proportion, to fill up the emptied space beneath, and so fall lower down than the rest of the Earth. And for the same reason, or others like it, many places in the Sea may be exceeding deep, and seem to go down into a perfect Abyss as it were, or a bottomless profundity.

And we must note, that though but only part of the Earth be Mountainous, yet little or none of it is exactly level; as being every where heaved up by the forementioned Causes, more or less. And therefore the smoothest Plains, that appear to the eye to be very even, are not really so. Only this we may observe, concerning them, That when

 Horse-men

Horse-men travel over them, the Ground being struck with the Feet of the Beasts, yields a kind of Sound. Which shews that the Earth in those Plains, is much in that Posture, into which the Sun and Vapours did at first raise it; loose, that is, and porous, and somewhat hollow. Whereas amongst Hills and Dales it yields no such noise, when beaten with such Tramplings. And the reason is clear, because it being flung up, and fallen down, and altered and transposed by eructations and sinkings; it has so been driven closer and made more compact.

And then as to *Maritime* Hills, or those near the Sea, when the Ground was crushed down by the hand of OMNIPOTENCE, to make a Receptacle for the Water; it is easie to conceive how they should fly up at the sides of Seas, or not far from them. As also how Hills should be highest in those Countries, about which Seas are deepest. For the Ground in the adjacent, or not far distant Seas, being sunk very low, and forced to give way very much; it might well crowd out and thrust up a great height, about the Shores, or in the adjoyning Regions. Nor is it to be thought, that when so great a part of the surface of the Earth was pressed down, that the Ground should struggle out at the Brinks of the Ocean only, and in some considerable distance from the Shores: much of it would recoil from under the compression in the Sea it self, and fly up irregularly in innumerable places, where it could best do it. And hence might come Banks in the Sea stretcht out (as Mountains are on the Land) to extraordinary lengths. As also Rocks, and Flats, and Shelves without number.

Nor

Nor muſt this be omitted, That all the Mountains of the Earth, if raiſed according to this Conjecture, will have no reaſon to hold proportion in bulk to the Cavity of the Ocean. A thing which the common *Hypotheſis* of their Formation implys, and which lies as a main Objection againſt it. For thus the In-land Mountains would not be made out of the Sea at all. Nor would the whole quantity of Earth, which at firſt filled up the Cavity of the Sea, be caſt out into the Maritime Hills; but moſt of it be ſqueezed and forced down deeper into the bowels of the Earth.

Thus alſo *Iſlands* might be made (to take a ſhort ſtep out of the way we are in;) I mean ſuch as are not of the largeſt ſize; whether they be diſtant from all *Continents*, as the *Canaries*, *Azores*, *Heſperides*, and others in the *Atlantic* Ocean; or ſuch as lie in whole Fries by the *Main-lands*-ſide, as they do in ſeveral places of the World. Though many of this latter ſort, might be raiſed out of Mud or Dirt, deſcending in great plenty out of Rivers. So were the *Echinades* in the *Ionian* Sea, juſt before the mouth of the River *Achelous*. Or elſe they might be made by the flowing of the Waters into the Sea, when they were firſt drawn off the ſurface of the Earth. For then they running furiouſly down into the Pit, which Providence had fitted and appointed for them, might wear away the ground about the Verge thereof; and eating into its *Superficies* by the violence of their courſe, might divide it into a multitude of little Apartments: which afterward when the Sea was filled, might be petty *Iſlands* about its Coaſts; as the *Philippines*, for inſtance, and others in the Oriental Seas, which ſtand in whole Sholes, even thouſands

thousands of them together against *China* and *India*.

Whirlepools also by the same means might be made in the Sea (as well as chanels for Rivers underground by land) for the Earth being pressed down deep in some places, and thereby forced to ascend in others; kind of arched Vaults might so be formed. Which leading out of one Sea (or one part of a Sea) into another; the waters flowing through them, cause those *voragines*, or Gulfs at the top where they enter their subterraneous Pipes or Passages. Many of which Gulfs are so strong, that they suck in and swallow up whatever comes into them.

But to return, we need no more wonder at the *Greatness* or *Number* of Mountains made (in this method) on the Earth; than at the Granulosity or ruggedness in the rind of an Orange. And as the Mountains in truth bear no more proportion to the Earth's Dimensions, than those little pimples do to the fruit we speak of; so they and In-land Mountains both, may proceed from Causes not *altogether* unlike. Though now those Causes as to the Earth are so debilitated and wasted, that they are unable to produce the like Effects. Particularly that flatuous Moisture, wherewith at first it did abound, and might be put into it on purpose to make it heave in *general* into necessary inequalities; and in *places* to ascend into mighty Hills; is spent and gone. And we have no more reason to expect that the Earth should ordinarily send forth Mountains now; than that a dead ripe Orange pluckt off the Tree, should break out into such Wheals or Wens, as we see upon some.

5. One

5. One argument for Mountains in the first World, is yet behind, which shall end this Chapter. *There were METALS in the World*; And these, as all know, are *now* found at the Roots of Mountains. And they being the places whence they are digged now, it is a shrewd presumption they ever lodg'd in the same. Indeed the very generating them in the exterior Region of the Earth, does necessarily suppose *cavities* in it. And Cavities under-ground, do as necessarily infer *inequalities* above it. And here the *Theory* will receive another wound (perhaps an incurable one) in its *Hypothesis*. I mean where it makes the Antediluvian Earth * *all smooth and even without Mountains*; *all solid (to the Abyss) without caves or holes.*

* *Totam levem & æqualem sine montibus; totam solidam (usque ad Abyssum) sine cavernis.* Pag. 145.

But therefore to shun this great inconvenience, it fairly consents to the abolishing of Metals out of the first state of Nature. † *Some moreover add to what has been said, that in the first nature there were no Minerals or Metals: who according to our* Hypothesis, *I think, want not their Reasons.* But this is out of the Frying-Pan, into the Fire. For thus the *Fidelity* of *Moses* is assaulted, and another intolerable affront put upon the HOLY GHOST. For do not both inform us, That the City *Enoch* was built, and the Ark prepared before the Flood? But how cloud *either* be done without Iron Tools? Some Barbarous people, I have been told, do strange Feats in way of Architecture, by sharp stones: But the *Theory* allows not so much as * *greater loose stones*, or *rough Pebbles* in the primitive Earth. So that if they had not Instruments of Iron, the Men of that Age,

† *Dictis præterea superaddunt non-nulli, in primâ naturâ nulla fuisse —— Mineralia aut Metalla; quibus non deesse suas rationes existimo secundum Hypothesin nostram.* Pag. 154.

* Eng. p. 243.

Age, could never have compassed the Works aforesaid. Yet all such Instruments are positively excluded by the *Theory*, in these words, † *Nor were there of old, Instruments belonging to War or BUILDINGS.* Nor need we wonder there should not, when there were no Materials whereof they could be made. Nor could there be such Materials, when the World afforded neither Mines nor Metals. Nor could the World afford either of them, when it was not possible the Earth should yield them. And that it was not possible for the Earth to yield them, the *Theory* again does implicitly affirm where it says that the first *World was* * *wholly* artificial, and that *the furniture* or *provision of things* which it had, *was not of such as were* bred, *but of such as were* made.

† *Neque essent olim quæ ad bellum spectant---- instrumenta---- neque quæ ad ædificia.* In Præfat. ad Lib. 2.

* *Totus ille mundus artificialis, & apparatus rerum non natarum sed factarum.* Ib.

But the worst is still behind. *Tubal-Cain*, as Heaven assures us, *was an* † *Instructer of every Artificer in Brass and Iron, Gen.* 4. 22. Yet the *Theorist* professeth (and that in the second publication of his *Hypothesis, after* he had had time to consider well) * *as for subterraneous things, Metals and Minerals, I believe they had none in the first Earth: and the happier they; no Gold, nor Silver, nor Courser Metalls.* But then how *Tubal-Cain* could learn his Trade himself, and teach it unto others; must be a Riddle too hard for *Oedipus* to untie. Or else, which is the very truth, this *Assertion* of the *Theory* must be notoriously false; and not only flatly, but loudly contradictory to the most express Word of the Infallible GOD.

† לטש, *acuens, i.e. acutè erudiens.*

* Pag. 244.

This alone (should all that has been said besides fail) is enough to blow up, and finally to explode this New *Hypothesis* of the Earth's Formation: I mean, as it shews its great incongruity not

not only to Scripture, but alſo to Philoſophy. For had the Earth been originally framed, as that teacheth it was; then grant there could have been a Metallic Region in that part of it under the Water; yet that Metals, or Matter for any one of them, ſhould ever have aſcended through the Abyſs, into the upper Cruſt of the Firſt Earth; would have been utterly impoſſible. And therefore that egregious Philoſopher *Des-Cartes* makes this the reaſon why Metals are not found in all places of the Earth; *quia per aquas evehi non poſſunt*; *becauſe they cannot be carried* or drawn *up through the* ſubterraneous *Waters. Princ. part.* 4. §. 73.

CHAP. XI.

1. *That there were* open Seas *before the Flood, made evident from Scripture.* 2. Such *Seas* neceſſary *then as* Receptacles *for* Great Fiſhes. 3. *The* Abyſs *being no* fit place *for them.* 4. *A farther* Confirmation *of open Seas.* 5. *An* Objection *againſt them,* anſwered. 6. Another *Objection* anſwered. 7. *A* Third anſwered.

1. HE that from the Clifts about it, or in ſailing through it, beholds and contemplates the Watry Ocean: That views it (ſo far as eyes and thoughts can reach) in the ſtatelineſs of its Depth and wide Expanſion: That conſiders what vaſt and numberleſs Rivers it continually drinks up, and yet is never the fuller for all theſe Acceſſions: How far it extends its ceruleous Arms, and how much it diſgorges at Millions of Mouths, and yet is never the emptier for all its profuſions: That ſees its inceſſant and unwearied Motions, and how it ebbs and flows with haughty and incontrollable Reciprocations: That obſerves how it ſurges with every Wind, and ſurlily ſwells upon every Storm; and lifting up tumid ſcornful Waves, foams as angry at its Diſturbance: That marks how it frets and rages in a Tempeſt, and rolls it ſelf up into liquid Mountains, as if it threatned to mingle Floods with the Clouds, or in a pang of Indignation to quench the Stars, or waſh down thoſe Lights hanged out by Heaven. He that gazeth on the ſpatious Seas, or revolves ſuch thoughts as theſe of it in his mind, would be amazed to think that ſo im-

menſe

menſe an Element was once lockt up in a Vault under Ground, and wonder where the Earth ſhould have Cellerage to hold it. He would ſcarce believe, that ſo proud, and ſtrong, and furious a Monſter, could be kept in Chains; or was ever ſo tame as to be coop'd up contentedly in a ſubterraneous Cave. He would hardly be perſwaded, that it could be made to hide its head in an hole beneath; and to lie quiet and ſtill in a nightſom Dungeon, where for many Ages it never ſaw the Sun.

But how odd and uncouth ſoever it may ſeem, yet thus it was, ſays this *Hypotheſis*. The ſame *Primary Aſſertion* of it, that ſays, *The Exterior face* of the firſt Earth, *was ſmooth and uniform, without Mountains*; ſays alſo, it was *without a Sea*. All that prodigious Maſs of Waters, which Imagination (as comprehenſive as it is) knows not well how to meaſure, was once ſhut up in an inviſible Cell; and being clapt under Hatches, lay *incognito* as long as the firſt World ſtood. Not a Drop of it appeared all that while, but what ſtrained forth by evaporation; or tranſpired through the Pores of the thick skin'd Earth, when by the heat of the Sun it was put into a ſweat. As for the main Body of the Waters, they lurked and hid themſelves in a ſecret *Grotto*; nor could they be brought to quit their latent Dwellings, or to look forth of their cloſe and dark Retirements; till the Roof of their Lodgings fell in upon them, and juſtled them out of their Manſions, to make room for it ſelf.

But againſt this, there lies the uſual Exception; namely, That it fights with the Holy Scripture. For *that* informs us, That when GOD made *Adam*, he gave him *Dominion over the fiſh of the Sea*. But according to this *Aſſertion* of the *Theory*, *Adam* never

ver ſaw the *Sea*, nor *one Fiſh* in it, all his life long, though it laſted well nigh a thouſand Years: and ſo, impoſſible it was, that he ſhould have or exerciſe ſuch a Dominion.

And it is farther conſiderable, That *Adam*'s Dominion over the Sea, was not only granted him by Patent from Heaven: but moreover was part of GOD's *Image* which was ſtamped on him. Whereinſoever the *whole* did conſiſt, this, I ſay, ſeems to have been *part* of the Impreſs. *For GOD ſaid, Let us make Man in our Image, after our likeneſs, and let him have dominion* over the fiſh of the ſea, *Gen.* 1. 26. And ſo, to ſhut up the Sea within the Earth till the Flood, is to deny to Men a part of that Empire, wherewith their Maker was pleaſed to inveſt them; and to deprive them of a piece of his glorious Image which he put upon them. For none could ſhare fully in the one or the other; but they who lived after the general Deluge.

If it be ſaid, That Men at length were made Lords of the Seas (as ſoon, that is, as they were open) and had power over the Fiſh therein; and ſo the word ſpoken by their Creator, was ſufficiently verified; and the Prerogative promiſed, amply conferred: I anſwer,

Firſt, The Divine Word was never made good to *Adam*, nor was that high Prerogative beſtowed on him. Yet he being the Head of Mankind, had reaſon to be inſtated in all the Privileges of Humane Nature, which GOD annexed to it, or ſettled upon it, as ſuch.

Secondly, *Adam*'s Off-ſpring (as many as lived and died before the Flood) did no more partake of this Priviledge, than he himſelf.

Thirdly,

Thirdly, The Rule or Dominion over Seas and Fish, intended for *Adam* and his Posterity; was immediately conveyed to them. Even at the same time that Dominion was given to them, over Fowle, and Cattle, and Creeping things. And therefore we find it transferred by the same Act, and in the very same form of Donation. Only Dominion over the Sea was *First* mentioned; which is no sign that it was *Last* to be attained. The Royal Charter, by which they claim and hold the Prerogative, testifies as much: it runs as followeth, *And GOD said, Let us make man in our image, after our likeness: and let them have dominion over the fish of the Sea, and over the fowl of the air, and over the cattle, and over all the earth, and over every creeping thing that creepeth upon the earth, Gen.* 1. 26. And again *ver.* 28. *have dominion over the fish of the sea, and over the fowl of the air, and over every living thing that moveth upon the earth.* So that admit nothing of GOD's Similitude imprinted upon Man, did consist in this Dominion; yet let any judge, whether GOD did not intend hereby to pass unto Men, as full a Dominion over the *Fish*; as over other Creatures. As also whether they were *presently* to have and exercise this Dominion; or to be suspended from it for above sixteen hundred Years. That they had Dominion over the Fowl, and over the Cattle, and over the Creeping things; from the beginning; I dare say the *Theorist* himself will not deny. And how then can he bar them from it over the *Fish* till after the Flood? especially it being the *first* thing in order in the Sacred Grant. Which Clause, had they been kept from the benefit of it so long; would not only have been quite eluded, but miserably inverted. For then instead of Mankind's

Mankind's having Dominion over the *Sea*; *that* would have had Dominion, and a most Tyrannical one too, over *them*. Insomuch that the very first time they should have seen it, it would have drowned them all, even a whole World of them, save eight Persons.

2. And that there should be *open* Seas even from the Creation; seems very necessary upon the *Fishes* account. For when GOD gave them the Blessing of Multiplication (endued them, that is, with Appetite of Generation, and Power of Propagating their kinds respectively) he commanded them to *fill* (not תהום רבה the *great Deep*, or Abyss, but) המים בימים, *the Waters in the SEAS*. And therefore *Seas* there must be in the beginning of things; else Fishes could not have replenish'd them with their Breed. And indeed some kind of Fishes there were, that could be no where conveniently, but in *Seas*; as being too big for Rivers. For on the same day that other Fishes were made, GOD created huge Whales also; passing the same Benediction upon *them*, as he did upon those. I confess, תנינם does signifie other Creatures, as well as *Whales*: but the word, denoting *them* amongst the rest, that will be enough for our purpose. For Animals they were of so vast Dimensions, that where could they harbour but in spacious Seas?

* De Animal. l. 9. c. 49. * *Ælian* reckons up several sorts of them: as the *Leo*, *Libella*, *Pardalis*, *Physalus*, *Pristes*, and *Maltha*. Which last he calls δυσανταγώνιστον θηρίον, a Creature hard to be conquered. To which he adds the *Aries*, most mischievous and dangerous to be seen. For when he appears a far off, he troubles the Sea, and makes it tempestuous. But then he notes withal, that these Fishes come

no:

not near to Shores or Shallows; but keep conſtantly in the *Deeps*. And the ſame * Author remembers, that *Theocles* ſpeaks of Whales τειήρων μείζονα, *bigger than Galleys of three banks of Oars on a ſide*. And that *Oneſicritus* and *Orthagoras* wrote of Whales about *India*, half a Furlong long, and of proportionable breadth; and ſo very ſtrong, that oftentimes, ὅταν ἀναφυσήσῃ τοῖς μυκτῆρσιν, *when they puffed with their Snouts*, they would ſpout up the Water at ſuch a rate; that the unexperienced would take the Seas to be toſſed with Whirl-winds. Nor need we wonder at the exceſſive ſize of theſe Whales, when *Pliny* † gives account, that King *Juba* (in Books ſent to *Claudius Cæſar*, touching the Hiſtory of *Arabia*) makes mention of ſome that were ſix hundred Feet long, and three hundred and ſixty, broad. And the ſame *Pliny* ſpeaks of *Balænæ* in the *Indian* Ocean, as long as four Acres of Ground. *Mercator* alſo in his Deſcription of *Iſland*, beſides other huge Fiſhes, tells of the *Royder*, an hundred and thirty Ells long. And of a great kind of Whale ſeldom ſeen, like an Iſland, for magnitude, rather than a Fiſh. As alſo of the *Stantus Valur*, which, when it ſhows it ſelf, ſeems an Iſland, for bigneſs, and overturns Ships with its Fins.

* Ibid. l. 17. c. 6.

† Nat. Hiſt. l. 32. c. 1.

Now where could Fiſhes of ſuch prodigious greatneſs, move and multiply, but in vaſt and open Seas? * *Am I a* Sea, *or a* Whale? ſaid *Job*. He put them together, as having ſpecial Relation to one another. And truly if in the beginning, there were ſuch monſtrous Whales, there muſt be Seas anſwerable to them. And that the Whales at firſt created, were as large as any, we need not queſtion. For as it became the ALMIGHTY to ſend forth *them* in their full perfection, as well as other Creatures: ſo

* Job 7. 12.

so to convince us that he did so, he bestowed the Epithet, הגדולים, upon them, calling them, *GREAT Whales*, *Gen.* 1. 21. So that could there have been Rivers, or Lakes in the new Earth sutable to *other* Fishes; yet these mighty ones would have been too big for them.

3. What remains therefore, but that the only place of aboad which (according to this new *Hypothesis*) can be allowed these Bulky Creatures (though it does not allot it to them) must be the subterraneous Abyss? And then the Waters in that Abyss (how improperly soever) must be the *Waters in the Seas*, wherein they were to live and multiply, according to the Divine Blessing and appointment.

But *that* Abyss (though of a meet capaciousness) could by no means have been a fit Dwelling for them, upon several accounts. For,

First, It would have been a place exceeding *Dark*, full of perpetual and blackest Midnight. Neither Sun, nor Moon, nor Stars, could ever have lookt into it; or darted so much as one bright Beam into the Pitchy Recesses of it. So that besides the loss they would have been at for Prey, how could they have seen to direct their Motions? having no manner of Light at any time to guide them? So that upon occasion, they must have run at tilt upon one another; and being inclosed between two Earths, would have been in danger of stranding themselves both above and below.

Secondly, It would have been a place as *close*, as it was dark. And therefore what shift should they have made for Air? I think I may say for Breath. For as for Whales and other Fishes that have Lungs;
† *Nat. Hist. l. 9. c. 7.* † *Pliny* says, *It is fully resolved by all Writers, that they*

they breathe. And his Opinion it is, That all Water-creatures do the ſame, after their manner. In proof of which, he offers ſeveral Arguments not to be deſpiſed. As their *Panting*, *Yawning*, *Hearing*, *Smelling*, &c. To which add their *Dying* upon being frozen up for any time. Or if they be alive, their greedy flying to any little hole made in the Ice, whereat the Air enters. But in the Abyſs they could have had neither Air nor Breath; and ſo for lack of the ſame, muſt all have been ſmothered.

Laſtly, It would have been a place as *Cold*, as it was dark and cloſe. For the ſame Cover of Earth (of unknown thickneſs) that would have hindred Light and Air from piercing into the Abyſs; muſt have kept out the Suns cheriſhing and benign Warmth too. So that could they have ſtruggled with, and overcome the *two firſt* Inconveniences; yet here they would have met with a *Third*, inſuperable. Could they have lived without Light and Breath; yet they could not have multiplied without the Influence of Heaven. The want of that, would have chil'd and quench'd the deſires of Procreation in them, and rendered them impotent that way. Thus, Winter, we ſee, is no ſeaſon for Production of Fiſhes; as being deſtitute of that quickning power and encouragement, which the Preſence of the Sun affords.

4. Farther yet. That there were Seas in the Beginning, even on the *Third* Day; we are taught, *Gen.* 1. 10. *GOD called the dry land, Earth, and the gathering together of the waters, called he, Seas.* And why ſhould they not be ſuch Seas as we have now? For we have no more grounds to think or ſay,

That the Waters there mentioned, were an invisible, potential, or proleptic Sea; than we have to imagine or affirm, that the dry Land there spoken of, was an invisible, potential, or proleptic Earth.

And that there were open Seas then, may be argued from the Waters we read of *under the firmament*, *Gen.* 1. 6, 7. *And GOD said, Let there be a firmament in the midst of the Waters, and let it divide the waters from the waters. And GOD made the firmament, and divided THE WATERS WHICH WERE UNDER THE FIRMAMENT, from the waters which were above the Firmament.* But had there been none but River-waters in the first World, and not such an open and huge Collection of Waters, as we now see: the Firmament could not so properly have been said, to *divide the waters from the waters.* For then it must rather have been in the midst betwixt the *Earth* and the Waters; and so must have divided the *Earth* from the Waters; the *Earth* which was under the Firmament, from the Waters above it. For as for the River-waters, *they* would have been too inconsiderable, to have had the Partition made by the Firmament, predicated of them in exclusion of the Earth, or in preference to it. It would have been as if the KING should have said, Let a Wall be built betwixt the *Thames* and the Conduits of *London*, to part them; without taking any notice at all of the City, which is infinitely more remarkable than the Conduits are.

But therefore the *Theory* presents us with a *new* Notion of the Firmament, and makes it to be quite another thing, than what it has always been said to be; namely, That *Cortex* or Outward Region of Earth, spread and founded upon the Abyss. And so ſo the Waters of the Abyſs under that Earth, muſt be the Waters under the Firmament. I cite but two Paragraphs to this purpoſe. *(a) Any one at the firſt view might be able to gueſs, that this exterior frame which GOD eſtabliſht upon the Abyſs, is to be underſtood by that Firmament, which GOD is ſaid to have eſtabliſht between the Waters below and above,* *Gen.* 1. 6. & 7. And again, *(b) As to the Firmament between the waters, it was a remarkable Phænomenon of the firſt Earth, or rather the firſt habitable Orb it ſelf, which every way encompaſſed and ſhut up the Abyſs; and ſo divided the Waters above, from thoſe below.*

(a) *Primo intuitu facilè quis ſuſpicari poſſet, hanc compagem exteriorem, quam Deus ſtabilivit ſuper faciem Abyſſi, intelligendam eſſe per illud Firmamentum, quod Deus dicitur ſtabiliiſſe inter aquas inferiores & ſuperiores,* Gen. 1. 6. & 7. Theor. p. 124.

(b) *Et quoad Firmamentum interaqueum inſigne erat Phænomenon telluris primigeniæ, vel potius ipſe orbis primus habitabilis, qui gyravit undique & concluſit———Abyſſum; atque ita Aquas ſuperiores ab inferioribus ſejunxit.* Ibid. pag. 254.

But this truly is ſo far from giving any ſatisfaction, that it will rather bring the whole *Hypotheſis* to confuſion: I mean, while thus it runs againſt Scripture again, and that moſt directly and ſhamefully. For the *(Firmamentum interaqueum)* Firmament that divided the Waters; was ſo far from being a *Frame* or *an Orb of* Earth, or *the firſt habitable Earth,* that (as the DIVINEST SPIRIT tells us) it was that wherein the Fowls were to fly, which yet were to *fly* above *the Earth, Gen.* 1. 20. Yea, in that very Verſe it is ſaid to be *the Firmament of* Heaven. And by GOD himſelf is ſtiled, *Heaven*; *GOD called the Firmament,* Heaven, *ver.* 8. Even that very Firmament which divided the Waters; as we learn from the two foregoing Verſes. And therefore *the waters under the Firmament*, in the ſeventh Verſe; are ſaid in the ninth Verſe, to be *the waters,* מתחת השמים *under the Heavens.*

Heavens. I confeſs, the *Theoriſt* twits us for underſtanding by the *Firmament* what we commonly do; calling it an (*a*) *Unphiloſophic* thing. But I forbear to retort. It is enough to ſhew that the advantage lies ſo much on our ſide, and that the ingenious Philoſopher is ſo utterly loſt in his Notion.

(*a*) *Auſim enim dicere quantum rerum Natura innoteſcit adhuc, Firmamentum Moſaicum prout Vulgò intelligitur eſſe prorſus ἀφιλόσοφον.* Ibid.

And ſince to make the Earth before the Flood, to be this Firmament, is ſo impoſſible; as being manifeſtly repugnant to the Truth of GOD: what remains but that it ſhould be that diaphanous *Expanſum* ſtretched out betwixt us and the Clouds? which as it is conſtituted of *Air* chiefly; ſo it is the place wherein Fowls do fly, according as Providence was pleaſed to appoint. And to ſeal up this for a certain truth, it is known that the *Hebrews* have no other word whereby to expreſs, *Air*, but שמים, or, רקיע *Heaven*, or *Firmament*. Only whereas *this Aereous* Expanſion extends from hence to the cloudy Regions (where are the Waters above the Firmament; and therefore are called * *Waters above the Heavens*) we muſt note that there is *another* Firmament mention'd by *Moſes*. I mean that *Expanſe* of indefinite vaſtneſs, wherein the Celeſtial Lights are fixed: for as we read, *Gen.* 1. 17. *GOD ſet them in the Firmament of Heaven.*

* Pſal. 148. 4.

But then this Aereous ſpace we ſpeak of, being the *true* Firmament; this proves there were open Seas at firſt. Elſe (as was ſaid before) this Firmament muſt have divided the Waters from the *Earth*; whoſe ſurface (bating a few Rivulets) would have been entire under it: but could not ſo properly have divided the Waters above the Firmament, from the Waters under it; becauſe the Waters under the Firmament would have been in no united Body, and of no join'd or continuous *Superficies*; but (to grant what

what the *Theory* ſuppoſeth) diſperſed in Rivers running on the Earth, which would have been one huge unbroken Continent. Yea, and in a conſiderable Tract of Ground around this Earth, there would not have been ſo much as one Rill of Water neither; even according to the *Theory* it ſelf, allowing its Hydrography.

5. But here we meet with Oppoſition upon ſeveral Accounts. As firſt, if Open Seas were the Waters *under* the Firmament, in the primitive State of things; then the *Clouds* muſt be the Waters *above* the Firmament: But againſt this it is objected thus: * *If nothing be underſtood by the Celeſtial Waters, or Waters above the Firmament, but the Clouds and the middle Region of the Air as it is at preſent: methinks that was no ſuch eminent and remarkable thing, as to deſerve a particular Commemoration by* Moſes, *in his ſix days work.* To which I take leave to anſwer, That the Clouds, how contemptible ſoever they may ſeem, are no whit unworthy to be ſpecified or remembred by that famous Writer in his *Coſmopæia*, or Story of the Worlds Creation. And this will appear if we rightly conſider but *two* things concerning them; Their *Dimenſions*, and their *Uſefulneſs*.

* Theor. pag. 234.

Firſt, Look to their *Dimenſions*. Who can tell what vaſt and mighty things they are? To what length and breadth do they ſtretch out themſelves? and how do they cover whole Kingdoms at once with their ſhady Canopies? And then they are of anſwerable thickneſs too. So that interpoſing betwixt the Sun and us, they oftentimes turn day into night almoſt, by intercepting his light. Which in the Holy Philoſophy, as an act of Providence, is thus

thus ascribed to GOD. *With clouds he covereth the light, and commandeth it by that which cometh betwixt, Job* 36. 32. Sometimes they mount up and fly aloft, as if they forgat or disdain'd the meanness of their Origin, and scorn'd to be thought of earthly extraction. Sometimes again they sink and stoop so low, as if they repented of their former proud aspirings, and did remorseful humble Penance for their high presumption. And though I may not say they weep to expiate their arrogance, or kiss the Earth with bedewed Cheeks in token of their Penitence; yet they often prostrate in the Dust, and sweep the very lowest grounds of all, with their misty foggy trains. One while they are spread thin and single over us; another while they are doubled, trebled, and strangely pil'd up or whelm'd one upon another: or else built with *Stories* as it were, and made into several Concamerations. And therefore they are said to be, עליותיו *His* (that is GOD's) *Chambers, Psal.* 104. 3, 13. *Now,* they look like Ridges of Hills in our Horizon: *anon,* like a Row or Chain of Rocks: and by and by they hang like pendulous Mountains, or swim like floating Islands in the Aiery Ocean. *Here,* they pour down abundance of *Rain*; and *there,* as much *Hail*: in one place, they scatter *Sleet*; in another, deep *Snow,* and that for many hundreds of Leagues together. To say nothing of those glorious things the *Rain-bow, Parelia's, Paraselenes,* &c. *Thunder* also is from the Clouds. And yet it is a thing so very considerable, that GOD himself calls it his VOICE, in the *Psalms*; yea, his *Mighty* VOICE, and also his *Glorious* or *Majestic* VOICE. So much Power, and Glory, and Majesty is there in it, that it strikes awe and terror into the hearts of the best,

as

as well as of the greateſt. And certainly (the righteous being bold as a Lion) it was a greater ſign of its Dreadfulneſs, that a good Man's Heart ſhould *tremble and be moved out of its place*, at this *VOICE of GOD's excellency*, *Job* 37. than that the *Roman* Emperor ſhould run under his Bed.

Thus the Clouds appear to be ſtrangely capacious Veſſels, or Store-Houſes rather of *Meteor* Proviſions. And yet (which is admirable) when they are never ſo large, and never ſo thick; never ſo full, and never ſo heavy; and (as one would think) ſhould load the Air with inconceivable gravitation: yet they do not fall down and cruſh us to pieces, or bury us alive under Mountains of Ice. No, they bear up as lightly, and drive on as ſwiftly through the yielding Sky, as if they had no kind of weightineſs in them. And to whatſoever Philoſophy may impute this (as to their being always in Motion; their being turgid with Vapours; to the thick conſiſtency of the Air under them, or the like) the thing is really and greatly to be wondered at. And therefore *Pliny* conſidering it, was ſtruck with Admiration, and cry'd out as in a pang of rapture or ſurprize; * *quid mirabilius aquis in cælo ſtantibus? What is more wonderful than the Waters ſtanding in the Air?* And well might he think it a marvellous *Phænomenon*, when *the Ballancings of the Clouds* are ſaid by the ALMIGHTY to be his *own* works; and not only ſo, but *the* **wundrous** *works of him*, as he is *perfect in knowledge*, Job 37. 16. In which regard, the Etymology of שמים *Heavens*, is not unfitly fetcht from, שאה, *to be aſtoniſh'd*, and מים. *Waters. Quòd ſtupendo modo aquæ illic ſuſpenſæ hæreant: Becauſe Waters hang there in an aſtoniſhing manner*, ſays *Buxtorf*. But why then ſhould they be thought ſo *deſpi-*

* *Nat. Hiſt. l.* 31. *c.* 4.

cable

cable by the *Theorist*, as to be *unworthy* of a particular commemoration by *Moses*?

They cannot be so, if, in the Second place, we consider their *Usefulness*. They are so far from being wholly *Superfluous*, or purely *Ornamental* things; that they are highly beneficial or *Useful*, *Three* ways. That is to say,

For the Earth,
For Mercy, and
For Correction.

* Chap 3. 7. 13. The Distinction is * *Job*'s; and therefore so authentic, that we need not scruple to go upon it.

First, They are *Useful* **for the Earth.** As they contribute greatly to the Preservation of it; to preserving it in a good and verdant State. If the same Great GOD whose Powerful Goodness brought the World into being, and fixt it in a Regular and Curious Order; did not by a wise and gracious manutenency (exerted chiefly in a well contrived Disposition and Concatenation of things, linkt to one another by a continued Chain of just Connexion and dependence) hold them fast together; they would soon shatter and dissolve into saddest Confusion. For though the Machin of the Universe, be as august, as it is immense; yet were it not for the accurate Symmetry of its parts, so skilfully fitted and connected amongst themselves: and for the mutual support which one piece derives and affords to another, by means of that necessary and elegant contexture, which runs through the whole habit or *Compages* thereof; it would immediately fall asunder, and rush into an heap of irreparable Ruines. Its Motions in some places would flag and faulter; and

and in others grown as much too fierce and violent: and ſo through unhappy deficiencies and redundancies of Motion (that commonly change and deſtroy Nature) poſt into innumerable Diſorders and Intanglements; and ſo become a moſt lamentably hampered thing, eternally devoid of all beauty and harmony.

And the very ſame would happen to the Earth in its Proportion. It is now a very goodly piece, and incomparably furniſht and adorned. There are few places in it but afford taking Proſpects, or preſent the eye with ſuch pretty Objects; that if the Beholders be not too incurious, they may well be affected with them. Herbs, Flowers, Trees, Fruits, Springs, Brooks, Rivers, *&c.* with what variety, and in what abundance does it ſend forth? But yet let the Clouds we ſpeak of, with-hold their moiſture but a few years; and what a rueful change would then appear? The choiceſt Grounds which now ſwell with Plenty, and luxuriate with fatneſs and pleaſing Gayeties; would be miſerably exhauſted, and their tempting amœnities turned into horridneſs. They would be quite deveſted of their florid attire, and of all their rich and gorgeous habiliments. Yea, not only their wanton gawdy Dreſſes, but even their coarſeſt and moſt ordinary Cloaths would be ſindg'd off their Backs: and being ſtript of their decent neceſſary Garments, would have nothing left to cover their nakedneſs.

We live in an Iſland, where (according to St. *Peter*'s phraſe) the Earth ſtands *in the Waters and out of the Waters*, more than in other places. Yet, as much Water as we have about us, ſhould the Clouds be unkind, and deny us their effuſions; to what grievous ſtraits ſhould we ſoon be reduced?

We may justly conclude so, from what has happen'd by some short Droughts amongst us; the effects of which are found upon Record in our *English* Chronicles. And if a little dry weather be intolerable to *us*, who dwell so near the Seas, and have *Neptune*'s Territories round about us; how extreamly pernicious must lasting Droughts be, to higher or more In-land Situations?

But therefore the *First Use* of the Clouds, is to keep the Earth in a Flourishing Condition. To temper the immoderate heat of the Sun, and to asswage his scorching fury. To moisten the Air, and keep it cool; and to cool the Earth by keeping it moist. That so once in a year at least it might put on its bravery, and be deckt and array'd in its prideless Gallantry; the Image of its native finery, and those higher glories, wherewith at first it was better beautified and imbellisht. And therefore when GOD brings the Clouds over it, to perform their work of natural Distillation; He is said to do it, לארצו, *for* His *Earth*; in the quoted Text. Because the Earth is *His*, and because it might continue to be *like* his; that is, Comely and Graceful. Whereas if Clouds by their Waters should not refresh it, in a short time it would scarce be fit to be owned for GOD's Earth. It would be so sear and bare, and barren and desolate; that it would hardly look like a piece of *His* Workmanship. Yea, so parched would it be, and so dry would it grow, and such heats would it conceive from the inflaming Sun; that it would be forced to anticipate its final Destiny, by burning, in good measure, before the Conflagration.

Secondly,

Secondly, The Clouds are *Useful* for Mercy. They do not keep the Earth from Desolation, and help to maintain it in a Condition good and flourishing, upon its *own* account only, but upon *ours* also. That so we may be fed, and cloathed, and furnished, with its valuable Products, and the fruits of its increase. The *Psalmist* (if we would read this in Holy Stile) expresses it thus: † *Thou visitest the earth and blessest it, Thou makest it very plenteous. Thou waterest her furrows, Thou sendest rain into the little valleys thereof; Thou makest it soft with the drops of rain, and blessest the increase of it. thou crownest the year with thy goodness, and thy clouds drop fatness. They shall drop upon the dwellings of the wilderness, and the little hills shall rejoice on every side. The folds shall be full of sheep, the valleys also shall stand so thick with corn, that they shall laugh and sing.* And in another place. * *He watereth the hills from above, the earth is filled with the fruit of thy works. He bringeth forth grass for the cattle, and green herb for the service of man. That he may bring food out of the earth, and wine that maketh glad the heart of man, and oyl to make him a chearful countenance, and bread to strengthen man's heart.*

† Psal. 65.9, 11, 12, 13, 14.

* Psal. 104. 13, 14, 15.

But this is not *all* the mercy which showrs down from the Clouds. They drop an higher mercy on us still: I mean as they are an Argument, and a mighty Argument, against the black and cursed sin of *Atheism*. For being notable Instruments of Divine Providence, they so bear witness in a powerful manner to the existence of a GOD. And therefore when as great a Disputant as ever entered the Christian Schools (except the adorable Master of them) would have reasoned Men into an acknowledgment of the True GOD; he argued from this very *Topic*, the *Clouds*; or (which is all one) from the Rain they afforded. Yea, he told them plainly, That GOD himself made use of it as an Evidence, to prove and attest his own being. *He left not himself without witness, in that he gave us* Rain, *Acts* 14. 17. And truly he that shall consider all the *Phænomena's* of this *Meteor*, and trace it along from its Rise or Generation, to its fall and profitable effects upon the Earth; will find it of singular force to evince a DEITY.

As for the Causes, Nature, and Qualities of *Rain*; the way of its Production, the manner of its Distillation, &c. the Apostle urged them not: he knew those things were too high for the Men of *Lystra*. But then he pressed them with the thing *another* way, more suitable to their Capacity; namely, As Rain was a means of the Earth's being fruitful. *He gave us* Rains *and* fruitful *seasons*. Pursue it but on this part, and how powerful an Argument or Testimony will it be, of the Existence of a GOD? I mean, as it will appear to be a wonderful Instrument, exactly fitted for its appointed work; and as it manifests a strange Providential Contrivance, in adapting it, in point of congruity and ability, to be the excellent Cause of such signal Effects.

For suppose the most understanding Man, as to that concern, in the whole World; had Woods and Nurseries, and Orchards and Gardens, and Fields and Pastures, to be watered: how would he chuse to have it done, so as it might be most for *their*, and consequently for *his own* advantage? Why, in the same way we shall find it done by the *Clouds*, only better: and indeed so much better, that it will be very

very hard, if not impoſſible, for *Art* to match it by any Invention. A certain Indication, that a more than ordinary or Humane Wiſdom, is intereſſed in the Affair. That the Clouds were made, that is, and alſo are managed by a GOD; whoſe Infinite Wiſdom indu'd them with that Nature, and placed them in that Order, and put them in that capacity of ſerving us as they do. So incomparably, that is, as no wit of Man can mend their Method. For let the skilfulleſt, I ſay, chuſe at what rate he would have his Grounds to be watered; and then ſee if the Clouds commonly come not up to his Rules, and exceed them too in what is fit to be done.

Firſt, We may be ſure he would appoint *the beſt kind of Water to be uſed.* And what Water ſo fit for all ſorts of Plants, as that which deſcends from the Clouds above? For conſidering how it is raiſed by the exhalative influence of the Sun, it can have nothing of ſaltneſs, acrimony, or deadneſs in it; nor yet of ſtarving thinneſs nor coldneſs neither; but muſt be as light, and unctuous and ſpirituous, as that Element, when ſimple, can well be; and by vertue of its ſutable qualities and conſiſtency, be moſt proper for invigorating the Seminals of all things. And then being drawn up from all parts of the Earth almoſt; as ſimple as it ſeems to be, there muſt needs be very great mixture in it. I mean, though it be all Water, yet it muſt be a Compound of all Waters as it were; as being an extract of all ſorts of moiſture that the Earth affords in its ſeveral Regions. Whence it follows, that all ſorts of Plants muſt find ſomething in it (it being originally in part derived perhaps from the Countries in which they grow) highly agreeable to themſelves; as conſiſting of of Particles fit to enter them, and easie to be turned into their substance. Which being suckt up by them, and drained by exquisite percolation through their fine digestive Pores; immediately becomes Sap (which is the Plantal Chyle or Blood) for their nourishment and accretion.

Secondly, Without question he would have these Waterings *seasonably performed.* And here the Clouds are most kind to Vegetables again; and by a regular method answer their necessities. For they yield both former and latter Rains. Such as may cherish them while they are young and make them grow; and strengthen them as they grow, and carry them on to perfection. Whereas if all these Rains should fall at first, the tender Springals would come to nothing; as being surfeited with too much moisture, and the principle of their Life irrecoverably chill'd, if not extinguisht. And if all should pour down upon them at last, the Showres would be to no purpose. For coming too late, they would be in vain: especially as to all Frugiferous things; which being shrunk and stunted with immoderate exiccation, would be unable to yield their kindly Products.

Thirdly, We need not doubt but he would have his Grounds watered *in a gentle manner.* And this, I may say, the Clouds do unimitably. Sometimes with dewy Mists; sometimes with greater, sometimes with lesser, commonly with soft and moderate Showres. Whereas should they discharge themselves in extravagant quantities; they would wash up the weaker, and beat down the stronger Plants; and by their too free and impetuous Defluxions, be extreamly injurious, if not fatal to both.

And

And can we think that what we have noted already, ſhould be done by meer *accident*? That the Regions above, which need them not, but are rather clogged and cumbred with them; ſhould draw up ſuch plenty of Waters for *us*, who cannot poſſibly ſubſiſt without them: and then ſend them down again of ſo elaborate a nature, at ſo ſeaſonable times, and in ſo ſutable meaſures: and all by caſual Oeconomy, and the conduct of blind and incertain Chance?

Fourthly, We may ground upon it that he would have theſe Waterings to be *conſtant*. Not only for two or three Months, or ſome few years; but ſo long as he lives at leaſt, to name no longer period. Nor are the Clouds deficient in this circumſtance neither. For as they have watered the Earth through all ages paſt, ſo they will do the ſame indefatigably for the future, even till the final Conſummation of all things. And though no *one Sett* of Clouds can ever be fixed or permanent, they being perpetually flitting and volant; yet as ſome fly from us, others ariſe; and ſo from new ſucceſſions of them, we have ſupplys of freſh Rain. And therefore albeit they are paſſant things, they leave very good and laſting effects of their tranſient fugitive preſence with us. And here the hand of Providence is viſible again. For put caſe that things by a fortuitous hit, had fallen luckily at firſt into that convenient poſture for Rain, in which now they ſtand (which would be moſt ſurpriſing to think;) yet that then they ſhould perſiſt of themſelves, ſo long, and ſteddily, and inalterably in the ſame; is not to be imagined. No, where the Wheel of Order runs on in ſo even, and withal in ſo laudable and holding a Courſe; 'tis a plain caſe that its Motions were

were derived from the impulse of Heaven; and are maintained by the help of a Divine Influence, or Providential Direction and Concurrence.

Fifthly, We may reasonably conclude, that he would appoint things to be watered *intermittingly.* Lest too much driness together, should injure them on the one side; or too much moisture prejudice or bane them on the other. Nor are the Clouds faulty in this piece of service, but perform it as it were, with a great deal of care and seeming Officiousness. For when they have poured out their kindness liberally on the Earth, they usually stop up their Bottles again; and by suspending their effusions promote its fruitfulness; as well as by sending them down upon it. For as Rains that are new and fresh from above, are most nourishing to Vegetables; so their intermissive descents make them to be more nutritive still. For then having drunk up and digested those that are past; they become more receptive of them that succeed. And so sucking in what is fit for their aliment with the more greediness; they disperse and concoct it with the more ease and speed.

And truly in the alternate vicissitudes of wet and dry weather, there is something, at times, most remarkably *Providential.* For when we have had sore and tedious Rains, for that very reason they should hold and increase; because Nature is prepared and inabled thereunto by abundance of Vapours. And when we have had a long and excessive drought, for the same reason it should continue; because Nature is fitted to carry it on; the parched ground affording fewer Exhalations, and there being a scarcity of matter out of which Rains should be made. Yet (as experience proves) it happens not thus, but on the contrary. For when Nature's Disposition in the case,

caſe, does ſenſibly ſtand one way; ſhe is turned about, and as it were againſt her ſeeming and ſet Inclinations, led into another. Which whiſpers and ſuggeſts to the thinking Man, that ſhe is certainly directed by an hand from above, and in theſe preterintentional and undeſigned changes (as we may call them) is over-ruled by a power ſuperior to her own; and alſo joined with ſuch Wiſdom, as orders her much better than ſhe could do her ſelf.

Laſtly, We may preſume that this Perſon would certainly have *all* his Grounds to be watered. That the one might be fruitful as well as the other, and all of them recompence the impartial care, with a general Fructification. And here the Clouds are not at all defective, but act their part in this neceſſary ſcene, moſt unexceptionably. For they ſpread out their melting dripping Wings even far and near; and oblige the whole Earth where it needs, (not to ſay where it has no want) with their moſt free and univerſal Disburſements. And truly were it not for their Waters ſo copiouſly ſhed down on the Earth, how miſerable would the Condition of Mankind be? But then when things are ſo well and happily ordered; as that a bleſſing ſo needful, is made ſo general; and is every where ſo common and eaſie to be had: what a bright beam of Conviction, as to the Being of a DEITY, darts forth and ſhines down from the blackeſt Clouds? For who but the Great GOD could have ſtretcht out ſuch Fountains in the Spatious Skies; and for the needs of Men throughout the World, have invented ſo adequate and incomparable Supplies?

Nor indeed are they Instruments of *Common* mercy only; but vehicles oft times of *Special* secular Blessing and Prosperity, to some Persons. So it appears by *Eliphaz's* Advertisement; who tells us, That GOD *giveth rain upon the earth, and sendeth waters upon the fields; to set up on high those that be low, Job* 5. 10, 11.

Thus we have seen how beneficial Rain from the Clouds is, as to the *Earth*. In which respect when GOD pleases to send it, 'tis said to come, *for HIS Earth*; as above noted. We have also seen how beneficial it is to *Men*, in watering their Possessions: and that in so singular a way, as the wisest could never have projected a better. In which regard it is said to come, *for mercy*. And so it does most signally, not only as it fills Men with *Temporal* good things for the use of their Bodies: but moreover as it is, or may be, a means of *Spiritual* Mercy to their Souls; in ministring an Universal Argument to Mankind, of the greatest *Truth*, and most necessary to be believed, of any in the World.

But then in the Third place (to pursue and fill up the Holy Man's Distinction) it is *Useful, for Correction*. As GOD is infinitely Good in Himself; so He alone is able to bring good out of evil. That's an Extract which none but the ALMIGHTY by a most Divine Chymistry peculiar to his MAJESTY, is able to make. And this he frequently does for those, in whose pure Affections he dwells and rules. *All things work together for good to them that love* GOD, *Rom.* 8. 28. But then there is another piece of his Character as true; That * *He will render recompence to his enemies*, and † *visit iniquities upon them that hate him*. And as he has numberless ways of doing this; so he often effects it by changing

* Isai. 66. 6.
† Exod. 20. 5.

changing things that are of neceſſary uſe, into fatal influence. Whereby he makes what is *naturally* good for Men; to be unto them *judicially* an * *occaſion of falling*. Moſt evident is this in the Inſtance of Rain. So needful is it, that we cannot ſubſiſt without it. Yet this very thing, GOD turns, when he pleaſes, into an heavy Rod; and by making it unſeaſonable or elſe exceſſive, chaſteneth his People ſorely with it. Yea, he imploys it not only as a Rod to chaſtiſe his Servants; but ſometimes as a Sword to cut off his Adverſaries; and an Inſtrument of Vengeance to ſweep away the ungodly in whole ſholes or multitudes. This was never ſo tragically apparent, as in *Noah*'s Flood; when a great part of that generall deſtructive, but deſerved Blow, which fell upon Mankind, was given by this Weapon. For by the Waters of Rain in conjunction with other Waters, a period was put to the firſt ſinful World, by a very juſt, though lamentable Cataſtrophe.

* Pſal. 69. 23.

And when the Clouds and their Rains, or *the waters above the firmament*, were ſo very conſiderable *in themſelves*; and withal ſo very *uſeful* in way of *Preſervation* to the *Earth*; and in way of *Mercy* and *Judgment*, as reaching out GOD's favour and ſeverity to the World: Why ſhould they not be worthy, and highly worthy, of *Moſes*'s notice in his Divine Coſmology? The Holy *Pſalmiſt* (who we are ſure ſpake by the ſame Spirit that *Moſes* did) looks upon the Clouds as mighty eminent and remarkable things. For as he makes them to be GOD's *Chariot*, *Pſal.* 104. 3. So in another place, he makes them notable Evidences of his Magnificence and Power; † *His worſhip and ſtrength is in the clouds.*

† Pſal. 68. 34.

 6. But

6. But against the Existence of *open Seas* at first, it is farther objected, thus, (*a*) *Nor are the Waters gathered* all *into the same place; for besides many salt Lakes, and some gulfs of the Sea perhaps heretofore impervious, the* Caspian *Sea, which is of the same origin and antiquity with the great Ocean, is far separate from it.*

(*a*) *Nec aqua congregantur omnes in eundem locum; nam præter multos lacus salsos, & aliquos forsan sinus Maris olim impervios, Mare* Caspium *quod ejusdem est originis & antiquitatis cum magno Oceano, ab eodem longe disjunctum est.* Theor. pag. 123.

To take off which, I answer; That *Moses* does not say, Let ALL the Waters be gathered into one place. Though if he had, the word A L L, in Scripture, is usually taken in a restrained sense, to signifie but a *Major* part: and so here it might have meant but the greater quantity of Waters. To give Proof of this out of the Writings of *Moses*. He tells us, * That *A L L the servants of* Pharaoh *went up with* Joseph *to bury his Father.* Yet we cannot think that the Court was quite empty at that time, and the King left wholly without Attendants. And therefore, ALL, there, must denote but *a great many.* So he delivered it as a Law to *Israel*; † *Three times in a year shall ALL thy males appear before the LORD thy GOD, in the place which he shall chuse.* And yet we know that some of them at those times must be decrepit, and some sick, and some unclean; and so unable to take such a Journey, and unfit to make such an appearance. And therefore by, A L L, here, can be intended but, *many*, neither; even as many as were capable of the performance, or qualified for it. And thus indeed ALL the Waters were gathered into one place. That is, the great quantity or main Body of them was so: as they were incorporate and united in the Ocean. Which whereever it diffuseth and insinuateth it self about the Earth,

* Gen. 50. 7.

† Deut. 16. 16.

Earth, is but one continued piece of Water, and ſo fills one continued ſpace with its huge *moles*. I ſpeak of a *partial*, and ſometimes a *ſecret* continuity; for it is not always open, viſible, and entire. And that the *Caſpian* Sea is a part of the great Sea, and holds a ſecret commerce with it under ground (as the dead Sea, or Lake *Aſphaltites*, is preſumed to do with the *Mediterranean*) is clear from hence; that it receives ſuch an abundance of Waters into it ſelf, and ſwells not with them. For though the Stream of *Volga* (which is thought to afford Waters enough in a Years time to drown the whole Earth) continually diſcharges it ſelf into the *Caſpian* Sea; it is never the fuller. And therefore the *Theory* need not have inſtanced in that Sea as a diſtinct and ſeparate Sea by it ſelf. Eſpecially when it allows it to * *have communication with the Ocean by Subterraneous paſſages*; whereby it is really, though not viſibly joined to it, and in ſome ſenſe, but one with it. And then as for other *Gulfs* and *Lakes*, that are diſtinct as to themſelves, and divided from the Ocean; how inconſiderable are they in proportion to it? But as ſo many Buckets-full to a large Pool. Yet ſhould the Waters run out of ſome huge Pool, and ſettle together elſewhere; as it might truly be ſaid of them then, that they are gathered together into one place, though many Buckets-full ſhould lodge in Plaſhes by the way: ſo the Waters in general, may rightly be affirmed by *Moſes*, to be gathered together into one place; though a Multitude of ſmall Receptacles, and the *Caſpian*, larger than the reſt, remain apart.

* *Per ductus ſubterraneos cum Oceano communicare.* Pag. 62.

7. But

7. But a Third Objection is yet to be removed (for I am willing to encounter all that are Material) which is this. If the Earth had *Open Seas* at first, dividing it into *Continents* and *Islands*, and interlacing and environing them, as now they do; how could the several parts thereof, so separate, be peopled with Men and stockt with Beasts? Or to use the words of the *Theory*, *(a) The propagating or conveying of Men and Animals into so many separate Worlds, would be difficult to explain.* I answer,

(a) *Propagatio & traductio hominum & animalium, in tot mundos separatos, difficilis esset explicatu.* Ibid. pag. 123.

First, It is as difficult to make out how the Earth should be peopled before the Flood, though the surface of it had been entire. I mean upon account of *that Torrid Zone* which the *Theory* supposeth to have been in it.

Secondly, Islands at first might be nothing so numerous as they are since. But as many of them were founded, as I may say, after the Earth; so many of them may be of later date than the Deluge. Which factitious or upstart Isles came into being Three ways.

Some were produced of an abundance of Filth, rolling down the Streams of Rivers, and running into the Sea and settling there. So were the *Echinades*, spoken of before. Concerning whose Production therefore, *Ovid* makes the River out of which they came, to speak thus,

——— ——— *Fluctus nosterque marisque*
Continuam deduxit humum, pariterque revellit
In totidem mediis (quòd cernis) Echinadas, *undis.*

Others

Others were thrust up in some Seas, and appeared on a sudden. Of this sort was *Rhodos* in the *Carpathian* Sea, an hundred and twenty Miles in compass; one of the ancient Academies of the *Roman* Monarchy. *Delos*, in the *Archipelago*, one of the Fifty three *Cyclades*. Remarkable for the Temple of *Apollo*; for most excellent Brass; and for the Fountain *Inopus*, which (as * *Pliny* affirms) rises and falls as the *Nile* does, and at the same times with it. *Alone*, hard by *Cyzicum*, and betwen *Lebedus* and *Teon*, Two Cities of *Ionia*. *Anaphe*, one of the Twelve *Sporades* (I think) or at least not far off them, as lying near to *Melos*, one of the chief of them. *Thera*, called also *Calliste*, where *Callimachus* the Poet was born, and whence they went who built *Cyrene*. It appeared first, in the fourth year of the hundred and thirty fifth *Olympiad*, as † *Pliny* relates; and from it was the Ilet *Therasia* broken off. *Hiera*, the same with *Automate*, which appeared about an hundred and thirty years after: *even in our time* (says the same *Pliny*) *upon the Eighth day before the* Ides *of* July, *when* M. Junius Syllanus, *and* Lucius Balbus *were Consuls.*

* *Nat. Hist. lib.* 2. *cap.* 103.

† Ibid. *l.* 2. *cap.* 87. *l.* 4. *c.* 12.

Other Islands again have been made by Disjunction from the Main-land. As some have been joined to Continents and become one with them; as *Æthusa* in the *Lybian* Sea, to *Mindus*; *Zephyria*, to *Halicarnassus* in *Caria*; *Narthecusa*, to *Parthenius*, a Promontory of *Arcadia*; *Hybanda*, to *Ionia*, and the like: so some on the contrary have been ravished or rent away from the firm Land. Thus *Prochyta* an Island in the *Tuscan* Sea, was raised not far from *Puteoli*: While a great Mountain in *Inarime*, falling by an Earthquake, poured forth that abundance of Earth of which it was composed.

posed. And so it carries the account of its Original in its (*a*) name; as (*b*) *Delos* also above mentioned does. *Cyprus*, a noted Island in the *Mediterranean*, was divided from *Syria*, says *Pliny*; whence it is now distant at least an hundred Miles. *Sicily*, from *Italy*. *Eubœa*, from *Bœotia*. *Besbycus*, from *Bithynia*. And as some have thought, *Britain*, from *France*. And truly if *Syria*, and *Cyprus*, which are now so remote from one another, were once united; this makes it the more probable, that *England* and *France* might (time out of mind) have been joined by an *Isthmus* or neck of Land.

(*a*) Ἀπὸ τοῦ περιέχειν. (*b*) *Quasi* Δῆλος.

Thirdly, It may be answered, That as Islands at first were not so *numerous*; so the bigger of them might not lie so far off from Continents as now they do: the Earth being since much eaten away by Waters, and so the distance betwixt them made much wider. Or if they did lie so far from the Main-land, yet the Inhabitants of such Lands, might advance into the distant Isles, by the help of some rude kind of Boats made of hollow Trees, or the like. Or if any were such out-liers, as that they did not designedly make towards them, or accidentally hit upon them; we may without inconvenience, grant them never to have been inhabited. And so we read of that *African* Island, St. *Thomas*, in the *Atlantic* Ocean, under the Æquinoctial; that at its first discovery (though since the Flood) it was unpeopled, and had nothing in it but Woods.

Lastly, I answer, As to the grand Continents of the Earth; *Europe*, *Africa*, and *Asia* (which are three of them) have known Inlets by Land, into one another. And, for ought we can tell, there may be Inlets out of *Asia* into *America*, in the Northern parts of

of them. But however we are ſure it is but a narrow Strait, that ſeparates the Kingdom of *Anian* from *Tartary*. And who can ſay but that before the Flood (and perhaps for a good while after it) there might be ſome Neck of Land coupling both together, and affording an eaſie Paſſage out of the one into the other; which may be ſince waſht down or ſwallowed up? For as the Earth does ſometimes gain ſtrangely upon the Water (witneſs the City of *Antioch* (to ſay nothing of *Ægypt*, the Bay of *Ambraſia*, the Flats of *Teuthrania*, and the now Meadowy level where *Mæander* runs, once belonging to *Neptune*'s Empire) which at firſt, ſays * *Pliny*, ſtood upon the Sea coaſt, but even in his time became an hundred and twenty five Miles diſtant from it:) ſo the Sea otherwhiles prevailes as much againſt the Land. Thus, the *Atlantis*, a vaſt Continent, bigger than *Libya* and all *Aſia*, (ſays † *Plato*) by a terrible Earthquake, laſting a day and a night, κατὰ τῆς θαλάττης δῦσα ἠφανίσθη, ſunk down into the Sea and diſappeared. And he that would ſee what the *Mediterranean* has devoured, let him but read the ſhort Ninetieth Chapter of the Second Book of *Pliny*'s Natural Hiſtory: where he gives a brief Account of what incroachments it has made in *Acarnania*, *Achaia*, in the *Propontis*, *Pontus*, &c. And when the Sea has been thus uſurping upon the Land, and has made violent breaches in ſeveral places; why might it not make a paſſage for its Waters between *Tartary* and *Anian*, though there was none at firſt? or why might not the ground ſink there by an Earthquake, or the like?

* *Nat. Hiſt. l. 6. c. 27.*

† *In Timæo.*

But grant the *Streight* we speak of, to have been ever the same that now it is: yet it will not be over difficult then neither, to conceive how *America* should come to be inhabited. As for *Quivira*, which lies right against *China*, and joines to *Anian*; that *that* was peopled out of *Tartary*, is not to be doubted. The course of Life which the *Americans* were found to lead thereabouts, does fairly show it; as being correspondent to that of the *Tartarian Hordes*, or *Scythian Nomades*. Only the Question will be, How the *Tartars* could swim over this Arm of the Sea, as having no shipping. For the resolving of which, we need but consider; that when the *Spaniards* grew acquainted with *America*, they found that the People upon the Coasts thereof, used little Boats made of the Trunks of Trees, hollowed (not by Iron Instruments, because they had none; but) by fire. Now grant but the *Tartars* (who dwelt upon the Coasts opposite to *Anian*) to have used the like; and how easily might they at times be accidentally hurried thither, in those sorry Sciphs, before they were aware? and so begin the Plantation of the *American* World. And then do but yield that the Inhabitants of *Tartary* before the Flood, were but as ingenious at making these *Canoes*, and as addicted to the use of them; and it might *then* be peopled the same way. Though what better Conveniencies for Transportation, the Antediluvian *Tartars* (as I may call them) might have; we cannot say. Nor can we hope ever certainly to understand, who were the *Aborigines*, or first Planters of the Post-diluvian *Americans*; or how they came into that spatious Tract of Ground, that half of the World of unknown extent, called the *West-Indies*. For the Natives being Strangers to Learning,

ing, have no Hiſtory amongſt them, or Records of their own Antiquities, that can make any tollerable diſcovery of this nature. As for their coming to the place where they built *Mexico*, under the Conduct of *Vitzliliputzli* their GOD, who went before them in an *Ark* (which account we have in the Story of the *Mexican* Kingdom, related out of their Memorials and Traditions) it is ſo general and obſcure; that no clear knowledge of their Beginning can be gathered out of it. I will only note therefore that as to the peopling of *America*, *Torniellus* is of our Opinion. For he ſays, * America *is a Continent with our World, or not very much disjoined from it, to which there might be paſſages by Sciphs or little Boats.*

* America *vel noſtro orbi Continens eſt, aut ab eo non valde disjuncta, ad quam ſcaphis aut parvis navigiis trajici poſſit.* Anno Mund. 1931. n. 49.

CHAP. XII.

1. *The* Scripture's *Silence touching the* Rain-bow, before *the Flood, does not argue its* non-appearance *till* after *it.* 2. *Its appearance from the* beginning, *no hindrance or diminution of its* Federal Significancy. 3. *But matter of* congruence *to* GOD's Method *of Proceeding in* other Cases. 4. Clouds *were extant before the Flood, and therefore the* Rain-bow *was so.* 5. *The* Conclusion *of this Chapter, relating to the* Two *foregoing ones also.*

1. I Should now have passed directly to the next *vital* or *primary* Assertion of the *Theory.* But there being no fitter place to do it in, I shall here bestow one short Chapter upon the *Rain-bow.* For that also is made use of collaterally, to support the *Hypothesis* we dispute against. Concerning it, GOD expresseth himself thus, * *I do set my Bow in the Cloud, and it shall be for a token of a covenant between me and the Earth. And it shall come to pass, when I bring a cloud over the Earth, that the Bow shall be seen in the cloud.* And because we heard nothing of it till *now*, 'tis presumed that it never appeared before. But by the same reason we may as well conclude, That a *Cloud* was never seen before neither; because here we find the *first* mention made of one. Which must needs be false, inasmuch as the Flood proceeded in great measure from violent Rains. And to suppose Rains without a Cloud, is the same absurdity, as to suppose Children without a Mother. And therefore our hearing nothing of

* Gen. 9. 13, 14.

of the Rainbow, till the Flood was paſt; is no good Argument that it *was* not, but that it was not *ſignificant*, before. And it is moſt like that we had not heard of it at laſt in the Sacred Volume (no more than we do of *Comets*, *Eclipſes*, &c.) if it had not been turn'd into a Pledge of Mercy to us. And in this ſenſe, as it was made a Symbol of the merciful Covenant; GOD might ſay emphatically, *I do ſet my bow in the Cloud*, when he ſtruck this Covenant with *Noah*. It was in the Cloud indeed before, but then it was only of *Nature's* ſetting there; and ſo it was but a meer *Cipher*, or at leaſt made no other Figure, but that of a bare *Phyſical Meteor*. But afterward when by Divine appointment, it became uſeful beyond its proper capacity, and never appeared without ſomething in it more taking than its colours; as being cloathed with a rare additional excellency, the new impoſition of a preternatural Signality: then, and ever ſince it might more juſtly and peculiarly be ſaid, to be ſet in the Cloud *by GOD*. For though it ſtands there ſtill in a natural and ordinary way; yet it ſerves to an extraordinary and ſupernatural end, and he it is that made it do ſo.

2. Nor would this ſignificancy which the ALMIGHTY put upon it, be at all impeached by its exiſting from the *Beginning*. For though it had appeared as commonly before the Deluge, as ever it did ſince; it would not for that have been the leſs authentick or aſſuring token of GOD's Covenant, or of his fidelity in keeping the ſame. For its force that way, depended not upon the Nature of the thing (applied to this ſymbolical uſe) but upon Heavens Inſtitution. So that if GOD had appointed

appointed the *Sun* to that use, he would have signified the same thing that the *Bow* does; though as all must grant, the *Bow* is the most fit Emblem of the two; and therefore it was chosen. For *that* never shows it self but in a Cloud, and with a Rain; both which were instrumental to the great Inundation: and so serve most properly to mind us of it, and from thence to pass our thoughts to GOD's Promise, of securing the World from such another.

Were it yielded therefore, that there never was a Rainbow, till that *Noah* saw after the Flood; what considerable Point would be gained as to G O Ds Design in exhibiting of it? What clearer Token would it have been of his Covenant? What stronger support of Mens confidence in it? the two principal Ends whereunto it was appointed. Why should *we* (for instance) that are now alive, be the more firmly perswaded of the Truth of G O D's Compact, or the more fully satisfied that he will surely stand to it? Indeed if it had been a *New* Apparition; by being so very fine and curious, it might have wrought prettily upon *Noah*'s Phancy, and theirs who were with him. Especially it coming with such a Promise of Mercy, and finding them in the midst of such gastly ruines. But bating but this, that its *Newness* might sweetly affect its Beholders, making delightful and somewhat surprizing Impressions on their Minds, and raising in them little transports and wondrings; what great benefit could result from it? As to the Persons then in being, it would have been a most valid Ratification of the Divine Covenant, without its novity: as being turned into a Seal of an immutable Promise of Security against general Floods to come, made by that

that GOD who had justly delivered them, and that most miraculously, from one lately past. And as to after Generations, it must be all one to them, whether the Celestial bow was first exhibited since the Deluge, or before it. For as many as think aright concerning it, that it is an Effect of Natural Causes; and that the rorid Cloud, the opposite Sun, and the Eye of the Spectator, being rightly disposed, so as to make due Refractions and Reflections of the Sun-beams at requisite Angles, it must as necessarily appear as fire must burn: they cannot discern any shadow of reason, why it should not as well assure them the World should be drowned no more, had it been extant ever since the Creation; as if it had commenced its appearance at the drying up of the Flood. To be short; whenever it appears, it is because it cannot do otherwise: but when it does appear, it betokens the Earth's preservation from drowning, meerly upon the account of GOD's Ordinance, that so it should do. And therefore it might be as good a *Prognostic* or Token of the Worlds indemnity from a second Flood, though it had appeared in all Ages before; as if it had then shown it self *first*, when GOD was pleas'd to make it the Sign of his Covenant.

3. And indeed it is the way or method of Providence both in its *Penal* and *Propitious* Dispensations, to make known, and common, natural, and familiar things, Marks of his Displeasure, and Significations and Vehicles of its kindness and beneficence.

Thus,

Thus, as to Punishment, it was a piece of the Serpent's Curse, that he *should go upon his belly.* Yet it does not appear that this Malediction deprived him of Legs, or that he and his *Species* ever went erect. So it was made part of the Womans Sentence, that her Husband should *rule over her.* Yet her Condition before, was a State of Subjection to him, as intimated to proceed from the order of her * Creation. So that what stands clearly imputed to her sin, and seems to be the plain Consequent of her Guilt, and the effect of her Doom; was antecedent to the same, and the Lot of her Innocence. Only Circumstances were altered, and what was sweet and easie as Nature at first; was unhappily changed into trouble and penance, in the issue of things.

* 1 Tim. 2. 13.

And the like is observable in the Oeconomy of Mercy. I mean, in the very conduct of Religion it self, and that in the sublimest Mysteries thereof. For the Evangelical Sacraments were instituted in *Water, Bread,* and *Wine,* for sealing and ratifying a far nobler Covenant, than that betwixt GOD and the Patriarch *Noah,* extending to his Posterity, and all living Creatures. And yet these were common Elements, and of ordinary use at all times. Only positive Commands and Divine Institution improved them into means of Christian Proselytism, and Communion with the DEITY.

And this makes it the more probable, that the Rainbow was an usual Meteor. Because then GOD in giving it to the World in Confirmation of his Promise or Pact; would have acted most consonantly to his other proceedings. Yea, even to his proceedings in the highest and holiest Solemnities of Religion; of nearest intercourse with His MAJESTY,

MAJESTY, and ſo of greateſt importance to Mankind. For there he has made the moſt common things, to be Signs and Myſtical Deferents of himſelf, and his Favours, to all worthy Partakers of them.

But *all ſuch inſtances*, ſays the *Theory*, *fall ſhort, and do not reach the caſe* before us. * For *a ſign confirmatory of a Promiſe, when there is ſomething affirmed* de futuro ——— *muſt indiſpenſably be ſomething* **New**. *Otherwiſe it cannot have the Nature, Vertue, and influence of a Sign.* And a little before, *Such Signs* ——— *muſt be ſome* **new** *appearance, and muſt thereby induce us to believe the effect* ——— *otherwiſe the pretended Sign is a meer Cypher and ſuperfluity.*

* Pag. 237. Read alſo pag. 236.

To which I anſwer, As to Signs given by GOD, to confirm his Promiſes ; he has taken a Latitude to himſelf in chuſing and appointing them. For,

Sometimes he has made things *new* and *ſtrange* to be Signs of this nature. Thus, his own Deafneſs and Dumbneſs, was to be a Sign to *Zachariah*, of his Promiſed Son. The Retrogradation of the Shadow on *Ahaz*'s Dial, was to be a Sign to *Hezekiah*, of his promiſed Recovery. And the Fleece expos'd to the Dew, firſt *wet*, and then *dry*, was to be a Sign to *Gideon*, of his promiſed Victory. But then,

Sometimes he has made things to be Signs, that on the other ſide are *common* and *uſual*. Thus the Fruit of a Tree growing in Paradiſe, was made a Sign of Man's Immortality, if he continued Obedient: and therefore it was called *The Tree of Life*, ſay many of the Learned. And ſhooting with Bow and Arrows upon the ground (than which nothing could be more ordinary) was made a Sign to *Joash* of his prevailing against the *Syrians*. And therefore when he shot, the Arrow was called *The Arrow of the LORD's deliverance from* Syria, 2 Kings 13. 17. Here was * *sagitta significans & promittens salutem*; or, in the *Theory*'s words, *a sign confirmatory of a Promise*, wherein *there was something affirmed* de futuro; but it had nothing *new* or *extraordinary* in it (the thing being most common and usual) save only that GOD by his Prophet, intimated its significancy that way. But had it not therefore *the nature, vertue, and influence of a Sign, whereby to induce the King to believe the effect? Was it a pretended sign* only, and *a meer Cypher and Superfluity*? The like may be said of the *Rainbow*. It was no *new appearance*, but a *Common Meteor* usually seen in the first World. But being stamped by GOD with a signality that way; it immediately put on the Nature, Virtue, and Influence of a confirmatory Sign; and became able to induce *Noah* and his Posterity, to believe that the Promise of the Earth's Preservation from future drowning, shall certainly be performed; according to the significancy wherewith it was marked, to ratifie that Promise.

* *Piscat. in loc.*

And no wonder that things extant and *common* in the World, should be made confirmatory Signs of GOD's Promises; when things transient and actually past long before (and so not to be taken cognoscence of but by remembrance) and things that never did or were to exist till long after his Promises should be accomplisht (and so as yet were no real things, and to be lookt at only with an eye of Faith) have been made such Signs. Of the first sort was * *the Sign of the Prophet* Jonas. Whose being vomited up of the Whale, after three days continuance in its Belly; was made a Sign of our SAVIOUR's

* S. Mat. 12. 39. 40.

SAVIOUR's rising from the dead, after his triduous abode in the Holy Sepulchre. Though *Jonah* was swallowed, and cast up by the Fish, near a thousand years before our LORD's interment and Resurrection. Of the latter sort was the miraculous Conception and Birth of the *Messiah*, which was made a Sign of safety promised to * *Ahaz*, against *Rezin* and *Pekah*; but was not brought to pass till above seven hundred and forty years after. Other instances of the same kind occur, *Exod.* 3. 12. *Isai.* 37. 30.

* Isai. 7. 14.

4. But we have a farther proof yet of the existence of the *Rainbow* before the Flood. And though it be but indirect and consequential, yet it may not want its weight. It is the *existence of Clouds then.* For if they were before the Deluge, as they are now; there were all causes needful for the production of the *Iris*: which could not but frequently conspire or fall in with one another, so as to paint that beautiful thing in the Heavens. And that there were Clouds in the first World, has been proved already by the same Arguments that evince there was a Sea and Mountains: for they imply and necessarily infer the being of Clouds. The Flood also was made of Rains in a great measure; and those Rains must descend from Clouds. And if Nature could produce Clouds *then*, she must be supposed to have done it long before; as being in a better capacity to effect it. For the Earth and Air could never be more hot and dry, than when the Deluge came. Scripture also gives countenance to this, that Clouds were extant from the beginning. *When he prepared the Heavens —— When he established the Clouds above,————When he ap-*

pointed the foundations of the Earth, Prov. 8. 27, 28, 29. Whence it appears, that GOD's establishing the Clouds, was contemporary with his preparing the Heavens, and appointing the foundations of the Earth. Indeed, שחקים, coming of, שחק, is πολύσημον, a word of various significations. It may signifie *Æther*, or *Air*, or *small Dust*. But then the *Heavens* and the *Earth* being both mentioned besides; *Æther*, *Air*, and *fine Dust*, must be comprized in *them*: which determines the word here to that sense in which it is rendred. And very properly; for besides that *Clouds* are said to be *the Dust of* GOD's *Feet*, Nah. 1. 3. the word in many places of the Holy Volume, does denote, *Clouds*; and that so directly and inavoidably, as it can be applied to no other sense. Nor may we forget that clear intimation, or evidence rather of the early existence of Clouds, *Gen*. 1. 7. Where the Waters above the Firmament, must be Waters *in the Clouds*, as has been already shewed. Even the *Theory* it self allows them not to be *Supercelestial* Waters. For as they are inconsistent with that *System* of the World which it goes upon; so it expresly disputes against them and * rejects them. And so what Waters else could they be, save those in the Clouds? Which grant them to have been, and how peculiarly were those *Clouds above* established (according to *Solomon*'s word) by GOD himself? when as yet there was no Sun to exhale them from the Earth.

* Pag. 16.

Let this be cast in as an Overplus; the *Rabbies* believed there were Rains in Paradise. (Though for some little time there might be none, *Gen*. 2. 5.) For when the LORD GOD put *Adam* into the Garden to dress it; they understand a kind of spiritual Cultivation of it, as he occasioned it to flourish by by his religious Performances. Particularly he presented, קרבנות, *Gifts* or *Oblations* for obtaining Rain and a right Influence of the Heavens upon it. Yet if there were Clouds and Rain, how necessarily must the glorious Bow appear, when Showres fell in a just position to the Sun?

5. To conclude this Chapter. If any thing in it will prove there was a *Rainbow*, or *Clouds* from the beginning; the same will prove *Mountains*, and *open Seas*. And if any thing in the Two precedent Chapters, will prove there were *Mountains* from the beginning; the same will prove *Seas and Clouds*. And if any thing will prove *open Seas* from the beginning; the same will prove *Clouds* and *Mountains*. For these three do mutually imply and depend on one another. And there being no good account of their *later* emergency into *Being*, but much to the contrary; they must in reason be thought to *coexist* from the *first*. And I remember, the learned and ingenious Dr. *More*, setting down some odd conceits of Philosophical Enthusiasts, puts this amongst the rest; *That there were no* Rainbows *before* Noah's *Flood*. *Discourse of Enthusiasm*, §. 44.

CHA

CHAP. XIII.

1. *The* Doctrine of Paradise, intelligible without the Theory. 2. Where *that Doctrine is* best taught. 3. What *it is*, *with* a brief Paraphrase *upon it.* 4. *It is* Clear *in it self, though* obscured *by Writers.* 5. Longævity *before the Flood, no* property *of Paradise; and might be the Priviledge but of* few. 6. *It could not be common to* all, *according to the* Theory.

1. THUS at length we come to Paradise. A place of greatest Fame, and of equal obscurity. For though touching it we hear very much, yet as to the site of it, we know but little. And to this Paradise the next *Vital* or primary *Assertion* of the *Theory* relates; which runs thus, *The Doctrine thereof cannot be understood, but upon supposition of the aforesaid Primitive Earth and its Properties.* But against this Assertion also we except; and do not doubt, but upon enquiry into the Doctrine of Paradise, to make it out, that it is Intelligible, without the help of the *Theory.* At least as intelligible *without* its *Hypothesis*, as it is *with* it; and that will be sufficient to our purpose.

2. Should any demand where *the Doctrine of Paradise* is best or most truly delivered; and what Writings contain the most authentic and credible account concerning it: whither could they be directed but to the Sacred Scriptures? For what there occurs in reference to the Paradisiacal State or Regions; may be firmly grounded upon as infallibly true.

true. Whereas what we meet with in ſome other Books, may be incertain, as written by Perſons of ſuſpected Credit. *Poets,* for inſtance, are by no means to be regarded in this matter. They are Men of wit and licentious fiction; and when they are ſtruck with their proper *Oeſtrum* or Rage, and grow warm in the vein of Romanceing; their Pens run on at a ſtrange rate, rangeing as far as quaint Phancy can carry them. But as to *them,* let thus much only be noted: That whereas the *Theoriſt* applys what they write of the *Golden Age,* to the Paradiſiacal State before the Flood; as if what they ſay, were ſome dark and imperfect Memoires of that: it might be diſputed (were it worth the while) whether they ſet not that Age juſt after the Flood; making *Noah* to be *Saturn,* and the principal Characters of the Golden Age, to fall in with ſuch things as happen'd in that Period. Several (of no contemptible learning) have been of that Opinion, and *Bochart* for one. As many as are diſpos'd to read what he wrote of this nature, may find it in the firſt and ſecond Chapters of his *Phaleg.* And if what Poets have delivered of the Golden Age, refers to times and things of a Poſt-diluvian Date; we have no manner of reaſon to regard them in the leaſt, as giving any light either into the Doctrine or State of Paradiſe.

Nor truly are *Fathers* (thoſe infinitely more excellent and ſolid Men) to be too much relied on in this caſe neither. I mean no farther than they are conſonant to the Oracles of Heaven, and write fairly after that inſpired Copy, which came down from thence. For though they be free from light and Poetic Figments, they are full of *Allegories* and high Rhetorications; and too *Hyperbolical* (erroneous ſome

of

of them) to be followed in all things. Thus when *Ephrem Syrus*, *Moses Bar Cephas*, *Bede*, *Strabus*, *Rabanus Maurus* and others, place Paradise near the Circle or Orb of the Moon; and St. *Basil* makes it τόπον ὑπερέχοντα τῆς ὅλης κτίσεως, *a place above the whole Creation*; and ἀνεπισκότιστον διὰ τὸ ὕψος, *obscur'd with no darkness by reason of height*; and the *Hebrew Masters* in general, will have it made before the World: how can these things be tolerably reconciled to a Terrestrial Paradise? And while some again (supposing the Ocean to incircle the Earth) place Paradise on the *other* Hemisphere; and then to bring Men into *this* after *Adam*'s Fall, will have the Ocean to be fordable, and People of that talness as to wade through it on foot: who can believe Paradise had such a Situation? Especially if we add, that other Doctors yet (of whom * *Bar Cephas* speaks) upon account of the Site of Paradise beyond the Ocean, held this Continent of ours was quite unpeopled and a kind of Desart till the Flood. And which still intangles things more and more, they generally concluded that the four Rivers of Paradise, were *Tigris*, *Euphrates*, *Ganges*, and *Nilus*; and that having their Spring-heads on the other side of the Sea, by a strange penetration or trajection, struck through the Earth, and brake out on this side of the same. To follow the Fathers here, can neither be safe nor easie. And better it would have been, if the *Theorist* had not gone so far after them; but instead of that, had kept to his word; † *We will never assert any thing upon the authority of the Ancients, which is not first proved by Reason, or warranted by Scripture.*

* De Parad. cap. 14.

† Pag. 4.

And

And therefore while Poets pursue the Golden Age in golden Dreams, and set it off in fine and extravagant strains: and the Fathers expatiate in too large and lofty *Encomiums* of Paradise, describing such rare and unaccountable Excellencies and *Phænomena's* of it, as it never had, but in their mistaken Idea's and Allusions: let us wisely attend to the Voice of GOD's SPIRIT in his unerring Word. So we shall learn what is fit and necessary concerning Paradise; and by keeping within the bounds of sober truth, shall never be cumbered with superfluous knowledge; nor be put to the trouble, first, of inventing humorous Notions, and then of defending them.

3. Now as to the Doctrine of Paradise, it is fully comprized (so far as we need to consider it) in the following Periods of Scripture. It will not be amiss to bestow a short Paraphrase on them.

Gen. 2. 8. *And the LORD GOD planted a garden eastward in Eden, and there he put the man whom he had formed.*

8. No sooner had GOD the LORD of all, brought Man into being, but in special respect and kindness to him, he assigned him his Dwelling in the sweetest Country of the whole Earth. In a place so ordered by the great care and wise contrivance of his Providence, that it abounded with delights; and for its exceeding pleasantness, was as the Garden of the World: situate in that Tract of Ground which is called *Eden*, and lies * Eastward from hence.

* Where *Moses* wrote the Description of it.

Mm Ver. 9.

9. And this Garden was most rarely furnisht. For besides Floriferous and Fruit-bearing Plants, the products of which respectively, were grateful to the Eye, and gustful to the palate, and useful for nourishment: there were two very extraordinary Trees. One, *the tree of life*. So called, because its Fruit (if eaten) would make a Man live very long upon Earth; and that without sickness, pain, or decay: or at least was a Symbol of Eternal Life, to be injoyed by him in a *better State*, upon condition of unsinning Obedience in *this*. The other, was *the tree of knowledge of good and evil*. Called by that name, because GOD had ordained that if *Adam* tasted its Fruit, he should presently know what Evil was, by a quick and sad sense of it; and the better know what Good was, by the lamentable loss of it. Both these Trees grew within the Garden.

Ver. 9. *And out of the ground made the LORD GOD to grow every tree that is pleasant to the sight, and good for food: the tree of life also in the midst of the garden, and the tree of knowledge of good and evil.*

10. And to the end that this Garden thus flourishing and fruitful, might so continue; a River was made to spring up in it, or to flow through it. Which how far soever it might run in one single Stream, and then divide into *two* or *three*; at last it fell into *four* branches or chanels.

Ver. 10. *And a River went out of Eden to water the garden, and from thence it was parted, and became into four heads.*

Ver. 11.

Which before they terminated or disburthen'd themſelves; as four ſeveral heads, were known by four diſtinct names, * after they had paſſed the Garden in one Current.

* *Rectè igitur* Lambertus Danæus *in Antiquitatibus ait, quatuor illa flumina fuiſſe unius & ejuſdem aquæ ſive fluvii ex* Edene *naſcentis divortia ſeu brachia. Et addit, ſontem iſtum & fluvium ex eo emanantem in* Edene *regione, antequam ſe in divortia illa quatuor diduceret, hortum illum terreſtrem irrigaſſe, & quidem totum adhuc & non diviſum; poſtquam autem totum hortum irrigaverat, tunc ſe inſra hortum in iſta quatuor flamina diviſiſſe.* Alſted. *Encyclop. l.* 20. *Hiſtor. c.* 11.

Ver. 11. *The name of the firſt is Piſon: that is it which compaſſeth the whole land of Havilah, where is gold.*

Ver. 11. As for the firſt of theſe Heads (for its fruitfulneſs in Fiſh, or the abundance of its Waters) it is called *Piſon*, and by * them that dwell near it, *Phaſis*, or *Paſitigris*. Which dividing it ſelf from *Tigris* (whereof this is the moſt Southern branch) about *Apamia*, runs along by the Land of *Havilah*: and parting that Land from the Country of *Suſiana*, it directs its courſe towards *Teredon*, and thereabouts empties it ſelf into the *Perſian* Gulph.

* *Paſitigrin incolæ vocant.* Curtius.

Ver. 12. *And the gold of that land is good: there is Bdellium, and the Onyx-ſtone.*

12. Of which Land of *Havilah* (whither *Saul* chaſed the *Amalekites*, 1 *Sam.* 15. 7.) it is memorable that there is Gold in it, and that Gold of an excellent ſort. It has alſo (the Tree, or, Gumm, or Pearl, called) *Bdellium*, and the *Onyx-ſtone*.

13. The second River or Branch (from its falling off or turning another way) is denominated *Gihon*. The very same that passing by * *Adiabene*, the most Northern Province of *Assyria*, compasseth the Land of *Chush*, or the Country of the *Asiatic Æthiopians*; that is, it glides along by it.

* See Mr. *Carver's* Discourse of the *Terrest. Paradise*.

Ver. 13. *And the name of the second river is Gihon: the same is it that compasseth the whole land of Æthiopia.*

14. The third Arm or River (from the sharpness of its Waters, or the swiftness of its Current) is called *Tigris*; whose course lies to the East of *Assyria*. And as for *Euphrates*, the fourth Head or Stream; it is so well known, that it need but be named.

Ver. 14. *And the name of the third river is Hiddekel: that is it which goeth toward the east of Assyria: and the fourth is Euphrates.*

15. And GOD directed *Adam* into the Garden of *Eden*, and placed him there; appointing him (as a piece of his recreation) to cultivate and order it.

Ver. 15. *And the LORD GOD took the man, and put him into the garden of Eden, to dress it, and to keep it.*

16. And as to the choice Fruits growing upon the several Trees in the Garden; GOD grudged him none of them; but gave him free leave and full power, to take when, and where, and as much as he pleased, even of the best of them.

Ver. 16. *And the LORD GOD commanded the man, saying, Of every tree of the garden thou mayest freely eat.*

Ver. 17.

Ver. 17. *But of the tree of knowledg of good and evil, thou ſhalt not eat of it: for in the day thou eateſt thereof, thou ſhalt ſurely die.*

17. Only he charged him very ſtrictly, that he ſhould not eat of the Fruit of the Tree of Knowledge: aſſuring him that if he did, it would coſt him dear; for he ſhould certainly loſe his Life thereby.

This is the Doctrine of Paradiſe, ſo far as at preſent we are concerned to look into it. What occurs in the Divine Story beſides, is rather of *Perſonal*, and *Moral*, than of *Local* Conſideration. It relates to *Adam* himſelf, rather than to the Paradiſiacal place of his reſidence. Only what we find in the Cloſe of the Third Chapter, muſt be taken in; which runs as followeth:

Ver. 22. *And the LORD GOD ſaid, Behold, the man is become like one of us, to know good and evil. And now leſt he put forth his hand, and take alſo of the tree of life, and eat, and live for ever:*

22. And now, my Angels, that Man has eaten of the forbidden Fruit, ye ſee how *wiſe* he is grown thereby. He has throughly tried the truth of that Promiſe the Serpent made him; and lo, how goodlily he has improved himſelf in GOD-like knowledge, which he aimed at, and thirſted after. Since he has been thus egregiouſly fooliſh, 'tis a thouſand to one, if we let him alone, but he'll fall into another piece of unreaſonable and undoing folly. And if he can but get to the Tree of Life, and taſte the Fruit of that, will

will presently conclude that his Life on Earth shall last for ever; and so neglect Repentance and Preparation for a better.

23. To prevent this therefore, GOD immediately turned him out of the Garden, into that place whereabouts he was created. And whereas in his Paradisiacal condition, he might have subsisted deliciously of what Nature yielded of it self; Now he was to live in a more painful manner, spending his strength in Tilling the Ground, that so it might afford wholsome sustenance for his Body, which was formed out of it.

Ver. 23. *Therefore the LORD GOD sent him forth from the Garden of Eden, to till the ground from whence he was taken.*

24. And being driven out of this pleasant Garden; to the intent he might never re-enter it more; GOD, by the ministery of Spirits, fired the Earth whereabouts his way of return lay into it. Which burning continually (as it does this day in many places, and as it did in * *Babylon* of old) was as effectual a means to keep *Adam* out; as if Providence had set a number of Cherubim to guard the passage leading thither, by brandishing flaming Swords.

Ver. 24. *So he drove out the man: and he placed at the east of the garden of Eden, cherubim, and a flaming sword, which turned every way, to keep the way of the tree of life.*

* *Campus est in* Babylonia *interdiu flagrans.* Plin. *Nat. Hist. l.* 2. *c.* 106.

4. Now

4. Now what is there in all this ſo difficult or abſtruſe, as not to be intelligible? The Doctrine of Paradiſe indeed has been ill handled, as well as ſome others: and has received great injury from ſuch as intended it nothing but kindneſs. Even eminent Writers, by exalting it too high, and inlargeing it too much; have unhappily obſcured it, and brought a wild Confuſion into it. Prepoſterouſly ſtriving to imbelliſh and improve it, they have mightily eclipſed and diſparaged it. Juſt as a true natural Beauty is ſpoiled by the addition of Artificial; and a lovely Viſage, made worſe by painting. But view it in its *ſacred* Pourtraicture, as *Scripture* has drawn or repreſented it; and ſo we ſhall have a fair Deſcription, a graphical or exact Delineation of it. So it will appear in its genuine Colours, and juſt Proportions; in its proper Features and due Complexion; and without all manner of diſgraceful Blemiſhes, or monſtrous Disfigurements. For,

Firſt, Here is nothing that turns the whole Story into Myſtery or Allegory. That makes Paradiſe it ſelf, to be the Soul; *Adam*, the Mind; *Eve*, the Senſes; the Serpent, Pleaſures; the four Rivers, four Cardinal Virtues, *&c.* Which is the way that *Origen*, *Philo*, and St. *Ambroſe* go. Nor,

Secondly, Is there any thing that intimates, the Garden of Paradiſe was the whole Earth. That the four Rivers mentioned in the Deſcription of it, had the Ocean for their Fountain. And that two of them, *viz.* *Piſon* and *Gihon*; were *Ganges* and *Nilus*: the one, running through *India*; and the other, through *Egypt*: according to the *Manichees*, *Becanus*, and *Noviomagus*. Nor yet,

Thirdly,

Thirdly, Does it mount Paradise up above the tops of the Hills, or assign it its Situation near the Moon, in an Earth different from ours: where *Bar Cephas*, *Bede*, and *Rabanus* set it.

These, we must confess, are things hard to be understood, and never to be made out. They contain in them unexplicable intricacies, and draw after them innumerable absurdities. Such as quite overthrow the truth of *Moses*'s Narrative, and so the Veracity of GOD himself. But therefore, as we see, they are no parts of the Doctrine of Paradise. That's entire without them, as Scripture delivers it; which makes them neither Essentials, nor Appendages of it. They are but διδασκαλίαι τῶν ἀνθρώπων,

* Colos. 2. 22. * *the Doctrines of Men*, according to the Apostle. Τερετισμὸς καὶ βαττολογίαι *fine and noisy words*, according to *Epiphanius*. And they that first spake them, being Persons of Fame, their Reputation gave credit to what they said; and their great Authority drawing others after them, they were followed by many, though themselves went not in the right way.

Lastly, Here is not the least touch, upon a *perpetual Æquinox*, or a *perpetual Spring*; or the *Pullulation*, or *Growing* of Animals out of the Earth;

† Pag. 176. which the *Theory* makes † Properties of Paradise. Nor is there a Syllable spoken, of *Adam*'s being formed at first on the other side of the World (that is, in the Southern Hemisphere) and then of his being transplanted hither. Nor is there any intimation, that the *Flaming Sword* was the *Torrid*

* Pag. 257. *Zone* (as the *Theory* * allows it to have been) but rather something that suggests the contrary. For Cherubim could no way be concerned in that; nor could it well be said to be *placed at the East of the Garden of* Eden, when it was placed round the Earth

Earth in way of Longitude. Nor do we hear a word, of Rivers ſpringing up on one ſide of the Earth , and then ſhooting through it to the other ſide, by deep and unintelligible trajections or tranſmeations. Thus a multitude of Difficulties would be ſtarted; and ſuch, as before we could run them down, would lead us a weary, yea, an endleſs Chaſe. But therefore theſe are no pieces of the true Doctrine of Paradiſe. That's clear, and obvious, and eaſie to be underſtood: at leaſt as eaſie as the *Theory*'s Doctrine concerning it is. For ſay that Two Branches of the Paradiſiacal Stream, be ſomewhat obſcure; yet we need not fear but they will as ſoon be ſet out to all Mens ſatisfaction; as the *other two* ſo well known, together with *Aſſyria*, &c. ſhall be found in the Southern Hemiſphere of the World, where, according to the *Theory*, Paradiſe was ſituate. Eſpecially when the upper Orb or Rim of the Earth fell into the Abyſs; and Rivers and Countries were all jumbled together in unſpeakable Confuſion.

5. As for *the Longævity of the Antediluvians*, that could be no Property or Adjunct of Paradiſe neither; inaſmuch as the common Parents of Mankind, were ſoon thrown out of it: and ſo the length of *their*, or of their *Childrens* lives, could not be owing to that State or Place, becauſe none of them lived and died therein. Indeed the *Theory* will have this *Longævity* to be a *Character* of the Firſt Earth, as *Paradiſiacal:* and * holds it was *common to good and bad, and laſted till the Deluge,* Mens *houſes of clay ſtanding eight or nine hundred years and upwards.* And though I will not poſitively deny this, That the People of the Firſt World

* Pag. 189.

N n did

did generally live to so wonderful an age (it being a received opinion) yet give me leave to ask; upon what good authority does it stand? The sacred Historians will hardly support it. For though he tells of † ten Men in a lineal descent that were long livers; yet this will be no conclusive Argument that all were so. For *they* were excellent Persons, and admirably useful upon several accounts. Besides founding and improving of Learning and Sciences; they were to instruct the World in Vertue and Goodness; to govern both in the Civil (it may be) and Ecclesiastical capacity; To countenance and propagate, as well as to defend the True Religion; to take care of the Worship, and promote the Kingdom and Interest of GOD; and to shame the looseness, and reform or restrain the lewdness of Men. And they being thus highly useful and needful; no wonder they were continued so long upon Earth. And thus we find *Noah* (it being requisite still in some measure upon the same accounts) living † three hundred and fifty years after the Flood, and reaching to nine hundred and fifty in all. And also * *Shem*, *Arphaxad*, *Salah*, and *Eber*, living (for the same reasons) much longer just after the Flood; than others did *then*, or have done *since*. Though we may say, of their long life, as *Rabbi Levi* (quoted by *Genebrard* in the first of his *Chronology*) did of the *Longævity* of the Antediluvian Patriarchs, that it was *opus Providentiæ, non Naturæ*; *the work of* Providence, *not of Nature*. Of such a miraculous providence as superintended the *Hebrews* in the Wilderness, and caused that their Cloaths by forty years wearing did not wax old. And then if we grant some of the prophane stock, (of the impious Race of *Cain*) to have lived

† Gen. 5.

† Gen. 9. 28, 29.

* Gen. 11.

lived as long (as the Ten Patriarchs before the Flood, and perhaps ſome few others not mentioned) by the ſame kind of Providence; for that they were exceding eminent in their ways, for very laudable and neceſſary things: (as *Cain* himſelf, for Husbandry and Architecture; *Jabal*, for Paſturage, and the ordering of Cattel; *Jubal*, for Muſick; *Tubal-Cain*, for Mechanics, and the like:) Grant, I ſay, but ſome of the degenerate Seed, of the worſer ſort of Men, to have lived a great while for perfecting the lower and leſſer Arts: as ſome of the Holy Seed and better ſort did, for carrying on things of an higher nature, and bigger concern: and poſſibly the Prerogative of Longævity, will be ſtretcht as far, as by the ſacred Records, it can upon *certain* Grounds, be extended. And though the *Theory* makes the Longævity we ſpeak of, common to all the Antediluvians; yet in the Sequel of this Chapter it will appear, that even according to the *Theory* it ſelf, it could not be a general thing.

But (in the mean time) if *Divine Story* proves not ſuch Longævity common to the Antediluvians; how ſhall *other Hiſtory* do it? The *Theoriſt* cites *Joſephus* as to this, and *he* brings in ſeveral Authors. What he ſays of the long life of them before the Deluge, I ſhall ſet down more fully than the *Theory* does. * *They being beloved of GOD, and newly created by him, uſing alſo a kind of nouriſhment agreeable to their nature, and proper to multiply their years; it is no abſurd thing to ſuppoſe that their years were of that continuance: conſidering that GOD gave them long life, to the end they ſhould teach Vertue, and ſhould conveniently practiſe thoſe things which they had invented in* Aſtronomy, *and by* Geometry; *the demonſtrations whereof they never had attained,*

* Antiq. l. 1. c. 4.

except they had lived at least six hundred years. For the great year is accomplished by that number of years: whereof all they bear me witness, who (either Greeks *or* Barbarians*) have written ancient Histories. For both* Manethon *(who wrote the History of* Egypt*) and* Berosus *(who registred the Acts and Affairs of the* Chaldeans*). together with* Molus, Hestiæus, *and* Hierom *of* Egypt *(who give an account of the* Phœnician Antiquities*) accord with me in that I have said.* Hesiodus *also,* Heccatæus, Hellanicus, *and* Acusilaus, Ephorus *and* Nicholaus *do declare, That they of the first World lived a thousand years. But let every Man judge of these things as he best liketh.* Where (to let pass other circumstances) let it be noted, that *Josephus* attributes long life, only to such as were *beloved of GOD.* and that to such ends as were now specified; that they might *teach Vertue,* and *use and improve* Astronomy *and* Geometry, wherein they could have attained to no considerable skill, without long life. And then as to the rest of those Authors he remembers; how could *they* understand the thing better than himself? For besides Scripture (which *Josephus* was much better acquainted with than *they*) what else could give them information in the case? And therefore their account, we know, is utterly false: for none of the first Worlds Ancients could ever reckon a thousand years compleat. Only some of them (in the sacred Register) came pretty near it; though most fell short of it by such a Period of time, as very few Lives comparatively now reach to. And that *Josephus* himself did not believe that all lived so long, as the Writers cited by him do mention; is plain from his shutting up the forequoted Chapter, with an expression showing diffidence

diffidence in himſelf, by allowing it to others. *Let every Man judge of theſe things as he thinks beſt.* Which I deſire may be noted the rather, becauſe there are few that write for this Longævity of the Prediluvians; but they ſtill quote this place in *Joſephus*, and back what they ſay, with the Authorities he brings. Yet we ſee, he is ſo far from being *poſitive*, in the matter, that he leaves People wholly to their own judgments about it. And as for Scripture, I ſay, he read and underſtood it as well as others; and could he have found good proofs of the Point there, he would doubtleſs have ſpoken more definitively of it. And the truth is, Scripture ſays not one word of *Cain*'s, or his Childrens living eight or nine hundred years. And therefore when we granted they might do ſo, it was no abſolutely neceſſary Conceſſion. But this is obſervable, that the Invention of *Manual* Arts, and ſuch things as might be carried on to good degrees of Perfection in a leſs ſpace of time; fell to *their* care and management. Whereas (according to the *Jewiſh* Hiſtorian) *Aſtronomy*, and *Geometry*, which could not be learnt but in longer Periods, were ſtudied by thoſe *Virtuoſo's*, who are upon Holy Record for *long-livers*. Which tacitly intimates that the *reaſon* of long life, and ſo *long life* it ſelf, was not common to *all*. And though *Moſes* remembers a few by name that lived ſo very long; yet this no more proves that *all* attained to the like age, before the Flood; than his ſaying * *there were Giants on the Earth in thoſe days*, does imply the whole Race of Mankind were ſuch. Yea, as his telling the World, there were ſome Giants *then*, does import that the reſt were otherwiſe: ſo his mentioning ſome ſo very long livers, may inſinuate that the

* Gen. 6. 4.

the rest were not so. Nor do we stand quite alone in this Opinion; For *Rabbi Moses* in his Book *de directione perplexorum*, as *Burgensis* cites him (*Addit.* 1. in *Gen.* 5. *apud* Lyr.) was of the same mind. And as he attributes length of life before the Flood, *miraculo divino, to divine miracle*; so he says, *diuturnitas fuit solùm in illis qui in sacra Scriptura nominantur, scilicet* Adam, Seth, Enos, *&c. non autem in aliis contemporaneis, qui non tam diuturnè vivebant; sed sicut post diluvium. Length of life was only in those, who are named in the Holy Scripture; that is to say,* Adam, Seth, Enos, *&c. but not in others their contemporaries; who did not live so long; but as* Men lived *after the Deluge*. *Burgensis* himself also (first, a most learned *Jew* of *Spain*, and then a famous *Christian* Doctor) seems to be of the same judgment with the *Rabbi*, in this matter.

But if at last it be urged, that the Authors aforesaid are too many and considerable, to have their Testimonies questioned or rejected; and that what they delivered of the Prædiluvians Longævity, must be true of them in *general*, they receiving the thing by authentic *Tradition*: let it be yielded. But then I must demand, and may be allowed to do so; How comes it to pass that *Tradition* is so *partial*, and not equally faithful as to other great concerns of the first World? Particularly, why does it not by the Pen of the same, or of other Writers, tell us explicitly of a *constant Æquinox*, and a *perpetual Spring*, as *Causes* of that Longævity? and not leave it to be imputed to *nourishment* and higher things; whither *Josephus*, we see, ascribes it. Why does it not tell us of a Sky without Clouds; and an Heaven without Rains; and an Earth without Seas,

Seas, and Mountains, *&c.* Surely if *Tradition* ſpake ſo loud in *one* caſe; and was ſo dumb or deeply ſilent in the *reſt*: this ſeems to evidence, that however there might be ſomewhat of truth in that *one Phænomenon*; there was none in the other.

6. Though truly that all the Prediluvians were ſuch long livers, cannot well be ſuppoſed for this reaſon. Becauſe then their Multitudes would have everlaid the Earth, and they would have wanted room wherein to ſubſiſt. For grant them to have multiplied but as Mankind did juſt after the Flood, or as the *Iſraelites* did in the Land of *Egypt*, or even as People do now adays; and where would there have been place convenient for them? And yet that they did increaſe at ſuch a rate, and faſter too, is but reaſonable to think; in regard humane Nature if ever it were ſtronger at any time than other, was ſo at firſt. To which add, that *Digamy* was in uſe before the Flood: and *Lamech* (one who was infamous for it) is ſaid by *Joſephus*, to have had ſeventy ſeven Children by his two Wives. Yea, perhaps Men were not only for *two*, but *many* Wives, *Geneſis* 6. 2. and Polygamy muſt contribute greatly to the encreaſe of Mankind.

But there needs no farther Proſecution of this. The *Theory* yields as much as we contend for, or can deſire in the Caſe: Though no more than what may be true, and ſo inavoidable. * *'Tis likely they were more fruitful in the firſt Ages of the World, than after the Flood; and they lived ſix, ſeven, eight, nine hundred years apiece, getting Sons and Daughters.* And again; † *If we allow the firſt Couple at the*

* Pag. 22.

† Pag. 23.

the end of one hundred years, or of the first Century, to have left ten pair of Breeders, which is an easie supposition, *there would arise from these in fifteen hundred years,* a greater number than the Earth was capable of ; *allowing every pair to multiply in the same decuple proportion the first Pair did.* So that if a *Supposition* (which (in the *Theorist*'s own judgment) is *easie*, may be but admitted (as why should it not?) either the Longævity of the Antediluvians must not be *universal*, or the Earth was *incapable* of the *number* of its Inhabitants. Nor could the Primitive Earth receive greater numbers of People than this. For grant it had no open Seas in it ; yet the *Middle* Regions of it being uninhabitable in regard of heat ; and the *Polar* ones upon account of Wet and Cold : both will be reduced to a pretty equal capaciousness. And should it be alleg'd, that the first Earth was bigger in Circumference, than this is ; that will be made good, by casting in on the present Earth's side, the sinking hollownesses and declivities of Valleys ; and the swelling protuberancies and gibbosities of Mountains ; neither of which the first Earth had.

Farther, if People before the Flood, were generally so long-liv'd, and this their Longævity proceeded from a perpetual Æquinox, and settled benign temperature of the Air, as the *Theory* holds ; then surely there would not have been that *difference* as to length of days amongst them, as we find there was. Thus, *Lamech*'s Age (as appears in the * Catalogue of long livers) was short of *Mathuselah*'s, near two hundred years. Whereas if the Cause of long life had been so uniform and steddy a thing, and so generally and equally

* Gen. 5.

equally influential upon all, as the ſuppoſed Æquinox; the Effect would have anſwered it: Longævity it ſelf would have been more regular, and not have admitted of ſo much diſparity. Though the truth is, ſuch an Æquinox, and ſuch an Earth as we have heard of, would rather help to ſhorten life (we may think) than draw it out to ſuch a length. For certain it is, that they muſt ſhut all Winds and Storms, and Clouds and Rains, and Thunders and Lightnings, out of the Firſt World. And what are theſe but *Criſes* of Nature, wherein thoſe malignities and noxious qualities which are lodged in her, and would corrupt her; ſuffer a Solution and are diſcharged? juſt as morbific humors in the Body, firſt ferment, and then are thrown off by proper Evacuations. But when there could be no Storms or Thunders, to put the Air into Motion, and to purge and clarifie it, that ſo it might continue pure and wholſome: it being always calm and too quieſcent (like ſtagnant Water) muſt needs putrifie, and contract ſuch foulneſs as would make it unhealthy, and apt to cauſe grievous Diſeaſes and Death. *Egypt* is almoſt in the pretended ſtate of the Primitive Earth. Situate between the ſecond and fifth Climates; its longeſt day not above thirteen hours and an half; has ſeldom any Rain, but is watered by a River. Yet how ſubject is *Cairo* to raging Plagues, and where are greater or oftener Mortalities than there?

I have only this to add here. If the Æquinox ſpoken of, were the cauſe of a *general* Longævity in the Prediluvian World; then other Animals would have lived as long proportionably, as Men. That is to ſay Lions, Bears, Wolves, Dogs, *&c.* And theſe multiplying five or ſix times (to say no more) as fast as Men; might have soon overpowered and destroyed them. Also Rats, Mice, Fowls, &c. multiplying (in that World) all the year round, and in far greater numbers than the Creatures aforesaid; would have destroyed Mankind another way: not by devouring *them*, but the *Fruits* of the Earth which they were to live upon. Especially when Men lived wholly on such Fruits (without eating Flesh) and had no such ways and instruments at first, of killing those Vermin, as now they have.

Nor did the Earth yield such plenty of Corn of its own accord, as to satisfie all granivorous Creatures, without preying upon Mens Corps. For upon Man's sin, * the ground was cursed. And upon that Malediction, it afforded not Corn without Tillage. For thence forward even *Adam* himself was to eat of it, בעצב, *in sorrow* (or *labour*) all *the days of his life*; nor could he have Bread, but at the price of his † *Sweat*. And if the first Men had no Bread-Corn, but what their industry fetch'd out of the Earth; how could they defend it against the swarms of devouring Creatures, increasing always upon them by numerous Procreations? Even barely to name all the sorts of them, that would be hurtful upon the account we speak of, and would unspeakably abound in a World that knows no season but Spring; is so great a Task, that I am willing to decline it. Yet that *other* Creatures did live proportionably as long as Mankind, the *Theory* owns; * where it makes the Longævity of *both* at once, a Third *Phænomenon* of Paradise and the first Ages.

* Gen. 3. 17.

† Gen. 3. 19.

* *Superest tertium Paradisi, & primorum seculorum Phænomenon, Longævitas hominum, &, ut par est credere, cæterorum animalium.* Pag. 160.

And

And which is very conſiderable alſo, it makes the firſt Earth the common Mother of all ſorts of Animals, which naturally bred them and brought them forth. Whence it muſt follow, that thoſe Terrigenous Creatures ſtrangely increaſing by ſpontaneous Births, would ſoon have filled the World, even this way alone (though they had not propagated their reſpective Kinds) with ſuch inconceiveable multitudes; as would have eaſily ſpoiled the Earth, and ruin'd Mankind. Who as they were made in the beginning but in *one* Pair; ſo they were capable comparatively but of a ſlow Multiplication. And ſo Beaſts, Fowls, Creeping Things, Inſects, and all manner of deadly and pernicious Creatures, would have poured in upon them in vaſt numbers, and with incredible forces; while they were unable to defend themſelves againſt them.

CHAP. XIV.

1. *The Flood could not be caused by* the Dissolution *of the* Earth, *and its* falling *into the* Abyss. 2. *For it would have been* inconsistent *with the* Description *of Paradise.* 3. *It would have* destroy'd *the* Ark. 4. *And have made the Earth of a* Form *different from what now it is of.* 5. *It would also have reduced it to a miserable* Barrenness. 6. *And have overturned the* Buildings *which outstood the Deluge.* 7. *And have rendred the* Covenant *which GOD made with* Noah, vain *and* insignificant.

Read the *Sixth Chapter* of the *first Book* of the *English Theory*.

1. LET us now go on to the next *Vital* or *Primary Assertion* of the *Theory*, which is this. *The Disruption and Fall of the Earth into the Abyss which lay under it, was that which made the Universal Deluge, and the Destruction of the old World.* For the vehement and piercing heat of the Sun, having parched and chapped the exterior Orb of Earth, and so greatly weakned it: and also having raised great store of Vapours out of the Deep within this Orb; their force at length grew to be such, that the Walls inclosing them being unable to hold them, the whole Fabric brake, being torn in pieces as it were with an Earthquake. At which time, the Fragments of that Orb of Earth, of several sizes, plunging into the Abyss in several Postures; by their weight, and greatness, and violent descent, caused such a rageing Tumult in the Waters, and put them into so fierce Commotions and furious Agitations, as made them boil and flow up

up above the tops of the new made Mountains; and ſo. cauſed the general Deluge. But againſt this we *Except* alſo, and ſay that the Flood could not be thus effected, for ſeveral reaſons.

2. Firſt, Becauſe it would be inconſiſtent with *Moſes*'s Deſcription of Paradiſe. What that Deſcription is, we have ſeen already; and 'tis done according to the proper Rules of *Topography*. For firſt, he marks it out by its *Quality*; a *Garden*. Then by its *name*; *Eden*. Then by its *Situation*; *Eaſtward*. Then by its *Inhabitant*; *Man*. Then by its *Furniture*; *every Tree pleaſant to the ſight, and good for Food.* And laſtly, by a *River* to Water it, which (riſing *in* it, or running *through*, or *by* it) did divide its ſtream into four Heads or Branches: all which, except one, are made to refer to ſome *Country* or other. Thus, *Piſon* is ſaid to compaſs the Land of *Havilah*: *Gihon*, the Land of *Cuſh*, or the *Aſian Æthiopia*: *Hiddekel*, to run towards the Eaſt of *Aſſyria*. But had the Earth been diſſolved to make the Flood; how could theſe Rivers, or how could theſe Countries, or any of either of them, exiſt in *Moſes*'s time; as being all ſwallowed up and for ever periſht in the fall of the Earth? And yet if they were not in being *then*, how could he deſcribe the Terreſtrial Paradiſe by them, as he does? And yet that they could have no being then, the *Theory* acknowledges in theſe words. * *'Tis true, if you admit our* Hypotheſis, *concerning the fraction and diſruption of the Earth at the Deluge, then we cannot expect to find Rivers now as they were before - ——— their chanels are all broke up.* But then if the *Hypotheſis* of the *Theory* were true, what meant *Moſes* to put theſe Rivers, or

* Pag. 252.

or any part of them, or any Countries near them, into the Topography of Paradise; when together with the Earth, they were all broke up and dissolved so long before?

To make the Argument as short as may be. In case these Rivers were *not* in the first World, it was impossible Paradise should be described by them. And if they *were* in *that* World, it was as impossible they should be in *this*. And so we have good evidence, that the general Flood could not be the Effect of the Earth's Dissolution. For if it were so, *Moses*'s Description of Paradise must be false. Which, to affirm, would be horrid Blasphemy, it being dictated by the HOLY GHOST.

Nor will it mend the matter here, to fall to *Cabbalizing*, or Expounding things *Mystically*. So we shall run upon the same Rock, and put hideous affront upon the Truth of GOD, by turning it into meer Figure and Falshood. Two eminent Fathers subscribe expresly to this. The first, *Epiphanius*, whose words are these. Ἐι τοίνυν ἐκ ἔςι Παράδεισος αἰσθητὸς, &c. * *If* Paradise *be not sensible, then there was no Fountain; if no Fountain, no River; if no River, no four Heads, no* Pison, *no* Gihon, *no* Tigris, *no* Euphrates; *no Fruit, no Fig-leaves, nor did* Eve *eat of the Tree, nor was there an* Adam, *nor are there Men; but the truth is a Fable, and all but* Allegories. The other Father is St. *Jerome*, who commenting on the fourth Verse of the first Chapter of *Daniel*; infers thus from it. † *Let their Dotage be silent, who seeking for shadows and images in the truth, do overthrow the*

* *Ancor. Sect.* 57.

† *Eorum deliramenta conticescant, qui umbras & imagines in veritate quaerentes, ipsam conantur evertere veritatem, ut flumina, & arbores, & Paradisum putent Allegoriae legibus se debere subruere.*

truth

truth it self, while they conceit that Rivers and Trees, and Paradise, ought to submitt themselves to the Rules of Allegory.

And here it may not be amiſs, to take notice how empty, and ſhallow, and extreamly trifling their reaſons are, that argue againſt a *Local* Paradiſe, and turn the Holy Story of it into Allegory. Let the Obſervation run but upon *one* Writer (who being as good as any that way, may ſerve inſtead of all the reſt) I mean *Philo Judæus.* (*a*) *Let no ſuch impiety invade our reaſon,* ſays he, *as to ſuppoſe that GOD tills the ground or plants a Paradiſe; inaſmuch as we might preſently doubt why he ſhould do it. For he could never thereby furniſh himſelf with pleaſant Manſions, Retirements, or Delights; nor could ſuch a fabuloſity ever enter into our mind. For the whole World could not be a worthy place or habitation for GOD; who indeed is a place to himſelf, and is full of, and ſufficient for himſelf.* Where the reaſon why there muſt be *no Material* Paradiſe, and why it is *impious* for us to think that GOD planted one; is, becauſe it would not be gratifying to him, and becauſe the whole World is not a fit habitation for him. (And therefore by the ſame reaſon there never was a World made neither.) As if Paradiſe had not been planted for *Man,* but GOD. And * elſewhere we find him harping upon the ſame ſtring, though it ſounds but harſhly. To take the Paradiſe planted by GOD, for a Garden of *Vines,* and *Olives, Apples, Pomgranates, and the like Trees, would be a groſs*

(*a*) Μὴ τοσαύτη κατάσχοι τὸν ἡμέτερον λογισμὸν ἀσέβεια, ὡς ὑπολαβεῖν ὅτι Θεὸς γεωπονεῖ καὶ φυτεύει Παράδεισον. Ἐπεὶ καὶ τίνος ἕνεκα, εὐθὺς διαπορήσομεν. Οὐ γὰρ ὅπως ἀναπαύλας ἐνδιαγώγους καὶ ἡδονὰς ἑαυτῷ πορίζῃ. Μηδὲ εἰς νοῦν ἔλθοι ποτὲ τὸν ἡμέτερον ἡ τοιαύτη μυθοποιΐα. Θεοῦ γὰρ οὐδὲ ὁ σύμπας κόσμος ἄξιον ἂν εἴη χωρίον καὶ ἐνδιαίτημα. Ἐπεὶ αὐτὸς ἑαυτῷ τόπος, καὶ αὐτὸς ἑαυτοῦ πλήρης καὶ ἱκανὸς ὁ Θεὸς, &c. Lib. leg. Allegor.

* *Lib. de Plant. No.*

gross and incurable folly. Τίνος γὰρ ἕνεκα εἴποι τις ἄν, &c. *For one might say; To what end was it? for a pleasant dwelling place? But then might not the whole World be thought the most contentful dwelling for* GOD *the Universal King?* And a little after; *Truly as* GOD *does not at all want other things, so neither* (Fruits for) *nourishment.* Where the main reason against a *Local* Paradise again, is (that which really is none) its *Uselessness* in reference to GOD. As if the design or end of a Paradise, had been to supply the necessities or conveniencies of the DEITY: and because GOD did not need it, and could receive no benefit by it; therefore it must be *folly* to think he planted it. But what was it that made so learned a Man to argue thus?

3. Secondly, The Dissolution of the Earth could not be the Cause of the general Flood; because it would have utterly destroyed *Noah*'s Ark, and all that were in it. For then that great and heavy Vessel, sinking with the Ground whereon it stood; must certainly have been staved all to pieces, if not overwhelm'd in the Ruines of the Earth. I know that in favour of this Ark, and for its Preservation; it is supposed that the * Abyss was not broken open till after the forty days Rain; and that those Rain-waters *might set it a-float*, and so prevent its ruinous Fall, by *keeping it from that impetuous shock, which it would have had if it had stood upon dry land when the Earth fell.* But this Supposition was noted above to be false, and must needs be so. For by the infallible Records we are assured, That the Fountains of the great Deep were broken up, and the Windows of Heaven opened in the *same day*, Gen. 7. 11. Yea, according to the order

* Theor. pag. 98

order of the Holy Words (if there be any Priority in those two Causes of the Deluge) the Disruption of the Abyss should precede; the breaking open of the Fountains being *first* mention'd. And so the Ark having no Water to Float on, must certainly have stood upon dry ground when the Earth fell. And consequently *the impetuous shock* spoken of, could by no means have been avoided; but must certainly have destroy'd the Ark, and all Creatures in it.

4. Thirdly, Had the Deluge been caused by the Earth's Dissolution, the Earth (or dry Land of this Terraqueous Globe) would in *likelihood* have been of another Figure than what now it bears. For under the *Ecliptic* (which in the Primitive Situation of the Earth (according to the *Theory*) was its Æquinoctial; and divided the Globe into two Hemispheres, as the Æquator does now) the dry ground is of *most* spatious extent and continuity. Even from the South-west parts of *Africa*, about *Guinea*; there is one entire Tract of *firm Land*, reaching as far as the *Persian* Gulf, and the *Arabian* Sea. That is, for the length of Seventy five Degrees, or between four and five thousand Miles. And then Eastward of that Sea, runs the main Land of *India*; which from the Western parts of it, to *Camboia*, in the East, is extended between two and three thousand Miles more. And yet it is all-a-long one continued Tract of Land, bating the *Sinus Gangeticus*, or Gulf of *Bengala*; which North-wards thrusts up but a little beyond the *Ecliptic* neither. And lastly, the same *Ecliptic* runs obliquely over almost the widest part of *America Peruana*; another piece of Ground three thousand

ſand Miles in breadth. So that the Earth ſeems to be too whole in its Æquinoctial Regions (I mean thoſe that were ſo before the Flood) to have been diſſolved to make the Deluge. For had it ſuffered ſuch a diſſolution; the *middle parts* of it falling in *firſt* (for ſome reaſons * before ſuggeſted) it ſeems probable that it ſhould have been more broken and ſhattered thereabouts than any where elſe; if not clean ſwallowed up: and ſo the Earth muſt have been of quite another ſhape than now it is. But this I ſpeak as a probable, rather than as a certain thing. Where grounds are but preſumptive and conjectural; Aſſertions built upon them, muſt not be poſitive and dogmatical.

* Chap. 9. Parag. 4.

5. Fourthly, Had the Earth been diſſolved to make the Flood; its Diſſolution would have brought it into a ſtate of moſt lamentable barrenneſs. For then the inward parts of it being turned outward; and the ſtarven Molds, and ſtony Materials in its Bowels, being made into its ſurface in a great meaſure: in all ſuch places, it would not only have been deſtitute of ſuch things, as ſhould have afforded nouriſhment both to Men and Beaſts; but moreover indiſpoſed *to*, and incapable *of* yielding them, for a long time. The Husbandman when he plows a little deeper than ordinary, and fetches up the *dead Soil*, as he terms it; it proves a great hindrance to his Crops. Yet what is *that* Soil, but part of what (upon the exterior Orb's tumbling into the Abyſs) muſt have been turned up by whole Countries at once? at leaſt in the Æquinoctial parts of the Earth, as being extreamly dried, and having all the heart or fatneſs ſuck'd out of it, by the ſcorching Sun. And where vaſt pieces of Earth ſank whole

as

as they were, and the ground alſo was of a richer nature (as retaining, we'll ſuppoſe, ſome of its native Oilineſs) yet *there* it muſt have been covered with an huge quantity of Mud, which would have made it barren by choking ſuch things as would have grown upon it. For the Waters below, being by the falling in of the Ground, expell'd from their aboad, and forced to fly up with unſpeakable violence; and then by reaſon of their plenty and gravity, deſcending with as much rage and force again; and ſtill as the Earth ſuffered more fractures, and plung'd into the Waters in more pieces, they feeling new commotions, and being huffed up and put into freſh eſtuations: by their riſing and falling, and working and beating furiouſly and inceſſantly; they muſt needs waſh and wear off a mighty deal of Earth from the fragments that dropped into the Deep. Which Earth being carried into all places, by the toſſing, rolling, turbulent Waters, and ſpread pretty thick upon the face of the Ground; and alſo incorporate with much other Filth; it could not but be occaſion of great barrenneſs to the Earth. For then when the Deluge ſettled and went off, that Filth could not but harden into a cruſt or cap upon the Earth's ſurface, very deſtructive to the Earth's fruitfulneſs. Eſpecially if we conſider, how long and diſmally the Ground was harraſs'd by the Flood, before it was incruſted. For, ſays the *Theory*, * *the Tumult of the Waters, and the extremity of the Deluge laſted for ſome Months.* And the fluctuations of the Waters being ſo boiſterous, and withal ſo laſting, they could not but waſh up, or kill moſt of the tenderer ſort of Plants, and many of the hardier and ſtronger ones too; yea, and perhaps rinſe off the

* Pag. 75.

top of the ground it ſelf, leaving it generally bare, and covering it in many places with ſtore of ſilty, ſandy, or gravelly ſtuff. So that the Earth being firſt made bare, and then overgrown with the Cruſt aforeſaid (which with the Sun and Wind would be baked on to it, and wax pretty ſtiff and hard about it) how could it at firſt have afforded ſuſtenance to the living Creatures? And therefore we read concerning † *Attica*, That *by reaſon of Mud and Slime which the Waters left upon the Earth, it was uninhabited two hundred years after* Ogyges's *Flood*. And that the whole Earth ſhould be in as bad a Condition after the general Flood; as *Attica* was after that Inundation which happened to it; we need not queſtion, if the *Theory* has hit upon the true Cauſe of the Deluge. So that however *Noah* and his Family might have made ſhift for Food (ſupporting themſelves by eating ſome of thoſe Creatures kept alive in the Ark, which G O D (at their going out of the ſame) gave them for meat, with a general Licence to eat Fleſh, *Gen.* 9. 3.) yet other Animals, for a time, would have been at a very great loſs for Nouriſhment.

† See Sir *W.* Raleigh's Hiſt. Book 1 c. 8. §. 11. † 5.

6. Again, had the Earth been drowned, by its being diſſolved and falling into the Abyſs; *all the Buildings erected before the Flood, would have been ſhaken down, or elſe overwhelmed.* Yet we read of ſome that outſtood the Flood, and were not demoliſht. Such were the Pillars of *Seth*, and the Cities *Henochia*, and *Joppa*. Touching which (to avoid quoting of ſeveral Authors) I ſhall only recite what I meet with in * one. *And for a more direct proof, that the Flood made no ſuch deſtroying alteration,*

* Sir *W.* Raleigh's Hiſt. of the World, l. 1. c. 3. § 5.

alteration, Josephus *avoweth that one of those Pillars erected by* Seth, *the third from* Adam, *was to be seen in his days; which Pillars were set up above fourteen hundred twenty and six years before the Flood, counting* Seth *to be an hundred years old at the erection of them; and* Josephus *himself to have lived some forty or fifty years after* CHRIST: *of whom although there be no cause to believe all that he wrote, yet that which he avoucheth of his own time, cannot (without great derogation) be called in question. And therefore possibly some foundation or ruine thereof might then be seen. Now that such Pillars were rear'd by* Seth, *all Antiquity hath avowed. It is also written in* Berosus *(to whom though I give little credit, yet I cannot condemn him in all) that the City of* Enoch *built by* Cain *about the Mountains of* Libanus, *was not defaced by length of time; yea, the ruines thereof* Annius *(who commented upon that fragment which was found) saith, were to be seen in his days, who lived in the Reign of* Ferdinand *and* Isabella *of* Castile. *And if these his words be not true, then was he exceeding impudent; for speaking of this City of* Enoch, *he concludeth in this sort;* Cujus maxima & ingentis molis fundamenta visuntur, & vocatur ab incolis regionis, Civitas *Cain*, ut nostri mercatores, & peregrini referunt. *The large foundations of which huge mass are to be seen, and it is called by the inhabitants of the Country, the City of* Cain, *as our Merchants and Strangers do report. It is also avowed by* Pomponius Mela *(to whom I give more credit in these things) that the City of* Joppa *was built before the Flood, over which* Cephas *was King: whose name, with his Brother* Phineas, *together with the grounds and principles of their Religion, was found graven upon certain Altars of*

Stone.

Stone. And it is not impossible that the ruines of the other City, called Enoch, *by* Annius, *might be seen, though founded in the first Age.* Solinus *also witnesseth concerning* Joppa, *that it* was *oppidum antiquissimum orbe toto, utpote ante inundationem terrarum conditum; the most ancient Town in the whole World, as being built before the Flood upon the Earth.*

Now if things were thus; that is to say, if a Pillar of *Seth*'s erecting (whereon was * ingraved the rules of Science) was standing after the Flood, in the Country of *Lycia*: If the City *Enoch* was so far from being ruined by the Deluge, that it was not defaced: If *Joppa* was so far from being swallowed up or made an heap of Rubbish, that the Altars in it were plainly discernible, and standing in such order, that the Inscriptions upon them were legible: then most certainly the Earth's Dissolution, and Fall into the Deep, could not cause the Flood. For then, suppose that the Ground had fallen but a Mile, or a Mile and a quarter downward; which we must grant it did at least (according to the heighth of the present Mountains, set at ten Furlongs, when carefully † measured by *Xenagoras* of old) and it would have given such a terrible jar or jounce, as would have shattered the abovesaid Structures all down, and laid them flat upon the Earth, if not sunk them into it.

* *Vid.* Joseph. *Ant. l.* 1. *c.* 3.

† Λέγουσιν οἱ γεωμετρικοὶ, μήτε ὄρους ὕψος μήτε βάθος θαλάσσης ὑπερβάλλειν δέκα σταδίους. Ὁ μέντοι Ξεναγόρας οὐ παρέργως, ἀλλὰ μεθόδῳ καὶ δι' ὀργάνων εἰληφέναι δοκεῖ τὴν μέτρησιν. Plut. *Paul. Æmil.*

And

And that which would have made it more difficult for them to have continued standing, was their Situation. For *Enoch* is said to be built *about the Mountains of* Libanus. But then about the Mountains the Waters would have been most irresistably violent, had the Flood proceeded from the Earth's Dissolution. So we are assured by the *Theory*. * *The pressure of a great mass of Earth falling into the Abyss ——— could not but impel the Water with so much strength, as would carry it up to a great height in the Air; and to the top of any thing that lay in its way, any eminency or high fragment whatsoever: and then rowling back again, it would sweep down with it whatsoever it rusht upon, Woods, Buildings, Living-creatures, and carry them all headlong into the great Gulf.* So that *Enoch* being situate about the Mountain *Libanus*, the very force of the Waters alone perhaps might have born it down. And then as to *Joppa*, I have some where read, That it is *oppidum monte situm* too, a Town situate on an Hill. Or if it be not, for certain it stands just upon the brink of the *Mediterranean* Sea: and so could never have escaped being overturn'd. For besides that it must have been shaken with the general fall of the Ground; it was placed just where the mighty Fragment, which dived into the *Mediterranean*, or made the bottom of it, was riven off; and so at the time of its hideous splitting off, the poor City must needs have suffered a very dismal Concussion. And the like may be said, in a good measure, of the Pillar of *Seth*, it standing not far from the Sea neither.

* Pag 75.

I know the very *Being* is questioned of *Seth*'s Pillar, *&c.* But what some doubt, others believe: and having *all Antiquity* (as the cited Historian says) on our side; we have ventured to put in this piece of Exception among others. *Valeat quantum valere potest.*

7. Lastly,

7. Lastly, Had the Dissolution of the Earth, been the Cause of the Deluge; *it would have made GOD's Covenant with* Noah, *a very vain and trifling thing.* Soon after the Flood was dried up, it pleased the great GOD to make an explicit and gracious Covenant with that Patriarch himself, and his Children; and in behalf of all Living Creatures then in being, or afterward to exist; that the World should be drown'd no more with such a general Deluge. And this Covenant he was pleased to ratifie with a remarkable Sign, that of the Rainbow; which was to be a lasting token of remembrance to HIM, as well as a Pledge of assurance to *us*. So we find *Gen.* 9. from the 8*th verse,* to the 17*th*. *And GOD spake unto Noah, and to his sons with him, saying, And I, behold, I establish my covenant with you, and with your seed after you: and with every living creature that is with you, of the fowl, of the cattle, and of every beast of the earth with you, from all that go out of the Ark, to every beast of the earth. And I will establish my covenant with you, neither shall all flesh be cut off any more, by the waters of a flood, neither shall there any more be a flood to destroy the earth. And GOD said, This is the token of the covenant which I make between me and you, and every living creature which is with you, for perpetual generations. I do set my bow in the cloud, and it shall be for a token of a covenant between me and the earth. And it shall come to pass, when I bring a cloud over the earth, that the bow shall be seen in the cloud. And I will remember my covenant, which is between me and you, and every living creature of all flesh: and the water shall no more become a flood to destroy all flesh. And the bow shall be in the cloud; and I will look upon it,* *that*

that I may remember the everlasting covenant between GOD and every living creature, of all flesh that is upon the earth. And GOD said unto Noah, This is the token of the covenant, which I have established between me and all flesh, that is upon the earth. Now if the Earth had been drowned the *Theory*'s way, what need of all this? Then it had been but GOD's telling *Noah*, how the Flood came; and that would have made him, and all his Posterity, both sensible and secure enough at once, that such *another* Flood could never happen. Yea, *that* scarce need to have been told him neither; inasmuch as the thing would have been throughly apparent, to them that lived in both the Worlds, from the great changes they must have observed: and so the Covenant would have been vain and useless. Yea, which is worse, it would have been perfect Mockery and Collusion; because then the Earth could not have been capable *of*, or liable *to*, such another Deluge So that GOD's covenanting not to drown it any more; would have been as if he should have covenanted that a thing impossible should not be done: that the Fire should not freeze, or the Sun shine darkness. For as neither Sun nor Fire can do such things, so long as they continue what they are; no more could the Earth be drowned a second time, so long as it continued a *dissolved* Earth. Yet that it *may* be delug'd again, is clear from GOD's *covenanting* that it shall not, and from the *Terms* of that Covenant. For the *Bow* in the Cloud is said to be a Token *of the Covenant.* And that when *that* Bow is *seen, GOD will* remember *his Covenant.* And that he *will look upon it*, to that very end, *that he* may remember *the everlasting Covenant.* Plainly intimating, that if that Covenant were not made; or being made, if by GOD it were not remembred;

 the

the Earth might again be drowned, with as universal and fatal a Flood as ever. But then if it *may* be so, from thence it will follow, that *Noah's* Flood could not be caused by the Earth's Dissolution. Because then Nature could no longer have been subject to a second Deluge, and GOD need not have covenanted to prevent it; His very doing it must have been a kind of imposing upon Men, as being but an ingaging to save them from an impossible evil; and to keep that sad Calamity off them, which nothing but miracle or his own Omnipotence was able to bring on. So that in fine, the case is come to this issue; Either that the Glorious GOD has done mighty unworthily (pardon the word) in making a Covenant, which has nothing but vanity and mockage in it; or else that the *Theory* determines falsely, in making the Deluge to flow from the Dissolution and Falling in of the Earth.

CHAP.

CHAP. XV.

1. *The Flood Explicable,* another way, as well *as by that in which the* Theory *goes.* 2. *What the* height *of its Waters might be,* viz. Fifteen Cubits *upon the surface of the Earth.* 3. *The* Probability *of the* Hypothesis *argued from* Scripture. 4. *What the* Fountains of the great Deep *were.* 5. *A* Second Argument *for the* Hypothesis, *from the* easie and sufficient Supply of Waters *to raise the Flood to* such *an* height. 6. *A* Third, *from its* agreeableness *with St.* Peter'*s* Account *of the Deluge.* 7. *A* Fourth, *from the* Habitableness *of the Earth, at the Flood's* going off. 8. *A* Fifth, *from its* Consistency *with* Geography.

1. WE are now come to the last *Vital,* or *Primary Assertion* of the *Theory,* which is this, *That neither* Noah'*s Flood, nor the present Form of the Earth, can be explained in any other method that is rational, nor by any other Causes that are intelligible.* That is, besides those which the *Theory* makes use of, or assigns. Now as to the present Form of the Earth, we have spoken something to that already. So that could but such an Explication of the Flood be given in, as would solve it, and the several *Phænomena's* of it, as *rationally* and *intelligibly* as the *Theory* does; *this* Assertion likewise would be sufficiently encountred in our way of *Excepting* against it. Let us therefore be allowed but *some* of that liberty which the *Theory* takes; that is, to make bold with Scripture a *little,* as *that* has done *a great deal*; and we'll try what may be done of this

nature. Not that I will be bound to defend what I say, as real and true; any more than to believe (what I cannot well endure to speak) that the Church of GOD has ever gone on in an irrational way of explaining the Deluge. Which yet she must needs have done, if there be no other *rational method* of explaining it, and no other *intelligible Causes* of it, than what the *Theory* has proposed.

2. We are now therefore attempting or roving at a *New Explication* of the Flood. And if in any thing it seems strange, let none wonder or be offended at it. We are only trying whether we can hit upon somewhat, that may be as rational and intelligible as to the matter in hand, as what the *Theory* offers; though it be as extravagant as that is. So that where we speak never so positively, still what we deliver, is to be lookt upon, not as an *Absolute*, but *Comparative* Hypothesis.

And first let us sound the Waters of the Flood. I mean by a true and infallible Plumb-line; even the same which *Moses* reaches out unto us, in the Seventh of *Genesis*. So we shall find there is a great mistake in the common *Hypothesis* touching their Depth. For whereas they have been supposed, to be fifteen Cubits higher than the highest Mountains; they were indeed but *fifteen Cubits high* in all, above the surface of the Earth. Not that the Waters were no where higher than just fifteen Cubits above the Ground: they might in most places be thirty, forty, or fifty Cubits high or higher. The reason is evident; because the surface of the Earth, were all its Hills gone, would be still uneven, and some parts of it considerably higher than

than others. Thus, * *Helvetia* is reckoned the highest Country in *Europe*. And in proof of as much, it sends forth four great Rivers into the four several Quarters of the *Europian* World. That is to say, the *Danube*, Eastward; the *Rosne*, Westward; the *Rhine*, Northward; and the *Po*, Southward. For though the Earth be a Globe, yet it is not one so true and exact, but were the Mountains taken off it, I say, it would still be rising or prominent in some places by the height of many Cubits, over what it is in others. At which rate, when the Flood ascended *fifteen Cubits* above the Earth where it is highest (which was the true height of the Flood) most of the surface of the Earth might be four or five times as deep under water. Thus, when *Switzerland* (suppose) was drowned to the height of *fifteen Cubits*; most of *Europe* might be drowned many times as high. And indeed that the Earth was uneven (as we have said) and much higher in some places than in others; cannot be doubted: it being but a wise and most necessary piece of Providence, that it should be so contrived. For otherwise spacious plain Countries (if habitable at all) would have yielded but incommodious Dwellings. I mean, because they must have been perfectly level, and so would have lacked devexities needful for Water-courses. For Rivers, we know, never flow but in way of *decurse* or running downward, off precipices, steepnesses, or declivities.

* Some think *Italy* the highest; because the ascent up the *Alpes* is very great on the *French* and *German* side; but the descent on the *Italian* side, inconsiderable.

3. This therefore we lay down as the Foundation of our *Hypothesis*, that the *highest* parts of the Earth, that is, of the *common surface* of it, were under Water but *fifteen Cubits* in depth: which would

would drown the reſt of its ſuperficies, very ſadly and ſufficiently. And this, I ſay, we learn from *Moſes*; who knowing it himſelf by Inſpiration, to inform us of as much, has committed it to writing, in the Seventh Chapter of *Geneſis*. For there we read at the eighteenth *Verſe*, That *the waters prevailed greatly upon the earth.* And at the nineteenth *Verſe*, that *the waters prevailed exceedingly upon the earth.* And how *greatly* and *exceedingly* did they prevail upon the Earth? *That* we have ſpecified in the twentieth *ver.* חמש עשרה אמה מלמעלה גברו המים **fifteen Cubits** *upward did the waters prevail.* What can be more clear or expreſs? They prevailed *fifteen Cubits* and no more. Fifteen Cubits *upward*; that is, upon the Earth. Upon which they are ſaid to prevail *greatly*, and to prevail *exceedingly* (in the two foregoing Verſes) that is upon the higheſt parts of its common ſurface.

And thus our *Suppoſition* ſtands ſupported by Divine Authority, as being founded upon Scripture. *That* tells us, as plainly as it can ſpeak, that the Waters prevailed but *fifteen Cubits* upon the Earth. (The cited Text, as a certain Plumb-line, ſhows them to have been no deeper, where the Earth boſoms out, and is moſt prominent.) And ſo it puts an uſeful key into our hands, to help us to unlock the myſtery of the Deluge, and to free the Doctrine of it from great difficulties and inconveniencies; which have run Men, it ſeems, upon *irrational* and *unintelligible* means and methods of explaining it.

4. Before we lay down any other Arguments in confirmation of the *Hypotheſis*; let us try if the light of Scripture, which ſhows the *Depth* of the Flood

Flood so plainly, will not also discover to us more clearly than yet has been done, what those *Fountains of the great Deep* were, which at the time of the Flood were *cleaved*, or * *broken up*. And truly this it seems to do very notably; giving us to understand, that they were but certain *Caverns*. Such Caverns, I mean, as were contained in Rocks and Mountains. And so the breaking up of the Fountains of *Tehom Rabbah*, or *the great Deep*, (which the *Theory* insists so much upon) was no more than the breaking up of such *Caverns*. This is evident from *Psal.* 78. 15. Where it is said, יבקע צרים, *He clave the Rocks* (the Rock *Rephidim*, and the Rock in *Cadesh*) *and gave them drink* בתהומות רבה, † *in Abyssis magnis, in the* great Deeps. That is, he gave them that for drink, which was in those *great Deeps* till he fetcht it out of them. And what *great Deeps* could *they* be, but great deep *Caverns* in the Rocks? and the better to evince, that the *breaking open* of the Fountains of the *great deep*, *Gen.* 7. 11. and the *cleaving* of those *Rocks* in the Wilderness, *Psal.* 78. 15. were, in effect, but the same thing: the same *Hebrew* * words are used in both places.

So, תהום רבה, are *Moses's* words: and בתהומות רבה, the *Psalmist's*.

But though these Caverns be called *Deeps*, we must not take them for profound places that went *down* into the Earth below the common surface of it: on the contrary, they were situate *above* it. And therefore the Waters issuing out of them, came *running down*. So we find in the next verse of the same *Psalm*, ויורד, *He caused them to* run down. And *Wisd.* 11. 4. the Water is said to be given *de petra altissima, from a most high Rock*. And *Gejerus* upon that

* Gen. 7. 11.

† So *Schindler*, *Buxtorf*, *Bithner*, &c. read the word: tho in a Copy by me, it is בתהומות.

* נבקעו, is *Moses's* word: יבקע, the *Psalmist's*.

that place in the 78*th Psalm*, does not only observe, that GOD made the Waters to descend, *ex petra præeminente*, *out of a very high Rock*; but also notes the reason why he did so, *ut origo aquarum omnibus pateret*; *that the source of the Waters might appear to all.* We cannot but remember likewise, that this water is said, 1 *Cor.* 10. 4. to follow the *Israelites.* Which speaks it to have had a fall from an *elevated* Situation. And indeed if it had not, it could not so well have run along with their Camp perhaps to *Cadesh*, where we next find them at a want for Water. Though if the Rock in *Rephidim* did supply them all-a-long in their intermediate marches and stages; we must needs conclude there were extraordinary accessions of Water into the *great Deeps*, or Caverns of it; out of which it flowed with so very plentiful and lasting Streams. The least that can be imagined, is, That they were so framed as to draw abundance of Vapours into themselves; which being dissolved in the Vaults within, from thence gushed out in a continued Torrent. Not unlike to the Waters in *Tenariff*, which every day pour down from a most high mountain; being generated (I conceive) of great store of Vapours which gather in some large hollownesses of the same, and through secret passages ascend to its Top. For on it there stands a certain Tree, continually covered with a Misty Cloud; which every day melting at Noon, discharges it self so copiously as to serve the whole *Island*; on which there never yet fell a showre, save that one which was forty days long.

I have

I have ſet down the ***high ſituation*** of theſe Caverns or Fountains, as foreſtalling an Objection, that might thus be made. If the great Deeps, whoſe Waters help'd to raiſe the Flood; were no other than Caverns; the Waters they afforded would contribute nothing to that uſe: for as ſoon as *they* had come out, ***others*** would have run into their places immediately, and ſo they had as good have kept in ſtill. But now theſe Caverns being of an eminent or ***raiſed*** ſite, the Waters they yielded towards the Flood, might help to ſwell it to its due pitch, according as we have ſet it; without any kind of danger, or indeed poſſibility, either of their own returning, or of others running into their room.

In caſe it be urged that Caverns, eſpecially Caverns ſo high ſituate, cannot properly be called *great DEEPS*: I anſwer, The HOLY GHOST has been pleaſed to give them that name, and his authority is not to be diſputed. So we find him ſtyling the *Red Sea*, תהום רבה. * ***the great Deep*** (as big a name as can be given to the vaſteſt profoundeſt *Ocean* now, and a bigger than was given to the *whole Maſs* of Waters at firſt, it being called but, *the Deep*, ſimply) which yet, for a Sea, was neither *Great*, nor *Deep*. Though thoſe Caverns which were opened at the Flood, might well be as *Deep*, as they were *Great*; meaſuring their Depths, from above, downward, towards the ſurface of the Earth.

* Iſai. 51.10.

And whereas the *Pſalmiſt* ſpeaks of *the great DEEPS*, as of *many*; and *Moſes* of *the great DEEP*, as but of *one*: this does not argue but the ſame thing might be meant by both. For as in Scripture, a *Plural* word, is ſometimes but of a *Singu-*

lar signification; (thus the Ark is said to rest upon the *Mountains* of *Ararat*, when it could rest but upon *one single* Mountain:) so a *Singular* word, does sometimes carry the force of a *Plural* one with it; (thus, השלו. *the Quail*, is put for the numberless multitude of them, *Exod.* 16. 13.) And therefore the different Numbers used by the Holy Writers in this Case, need not set them at variance, or imply that they intended different things. And then tho' *Moses* speaks of the great *Deep*, *singularly*, as but of, *one*; yet he speaks of *all the Fountains* of that Deep, as of, *many*: which makes the Expression somewhat more parallel to the *Psalmist's*, *great Deeps*. And then though the *Psalmist* puts the *Substantive*, *DEEPS*, in the *Plural* Number; yet he puts, *GREAT*, the *Adjective*, in the *Singular*: and so goes as far to meet *Moses* (as I may say) as *Moses* comes to meet him. And lastly, the *Septuagint* and *Vulgar* both, do render the *Psalmist's*, *DEEPS* in the *Singular* Number, Deep: as if it were no matter * *whether* Number were used.

* So, the *Red Sea*, which is called תהום the *Deep*, *Esai.* 51. 10. is said to be תהמות, *Deeps*, *Exod.* 15. 5. Yea, *Isaiah*, who in 51. of his Prophecy, calls it, תהום, in the *Singular* Number; in 63. Chap. 13. calls it, תהומות, in the *Plural*.

Should it be urged farther yet, that no such *Deeps* or *Caverns* are found in the Earth nowadays, and therefore it may be questioned, whether there ever were any or no: It might be answered. Though there are many of them, yet they may be of no easie discovery; as being inclosed with very thick Walls, and shut up within vastest and highest Mountains or Rocks. And truly so closely and strongly were they immur'd, in the Predilu-

vian

vian State; that had not ALMIGHTY GOD broken them up by his own Power (as he did those in *Rephidim*) they might have continued entire and undiscerned to this very day Though when by Omnipotence these mighty Cisterns of Nature were let go, and their Waters run out in a great measure; no wonder at all that the sides of many of them should cave in; making the Mountains or Rocks whereunto they belonged, very rough, and craggy, and deformed things; and scattering huge Stones, and such heaps of Rubbish whereabouts they fell, as might imitate the Ruines of a dissolved World; and show not only the Scars of a broken-fac'd Earth, but moreover (as one would think) the very Entrails of it strangely burst out, and as it were, torn and mangled all to pieces. And as a little marvel it is again, that the Crowns of several high Rocks and Hills, sinking right down into the Caverns beneath them, and being not able to fill them up; should leave huge Pans on their Tops respectively. While innumerable others yet, that were broached and well nigh drawn off at the Flood; have for many Ages stood dry and gaping : and have been Dens for wild Beasts; and sometimes Refuges, and sometimes, it may be, Habitations for Men; as being of very considerable Capacities. Of this sort, 'tis like, was that Cave in *Engedi*, which was able to receive *David* and his Six hundred Men: and for ought we know, might have held as many more. For these are said to * *remain in the sides of the Cave*, * 1 Sam.24.3. and were so well hidden; that King *Saul*, who was there at the same time, perceived not one of them. And that there were store of such Caves in *Palestine*, into which (in time of Invasion by Enemies, *&c.*)

Rr 2 the

the Inhabitants of the Country used to retire, even by whole Villages or Towns at once; is very well known. * *Josephus* makes mention of some of these Caves in high Rocks and Mountains; which being possessed by Robbers, King *Herod* was fain to let down armed Souldiers an unspeakable depth into them, in Chests with Iron Chains, to fight the Wretches in those their Fastnesses. † *Strabo* likewise reports, That towards *Arabia*, and *Iturea*, there are steep Mountains famous for deep Caves, one of which is able to receive Four thousand Men. Nor is it to be doubted but that in all rocky and mountainous Regions, there are plenty of most capacious Caverns. (a) The *Theory* it self allows them to be more common in such places, than elsewhere.

* *Antiq. l.* 14. *c.* 27.

† *Lib.* 16.

(a) *Sæpius enim in confiniis vel in ipsis visceribus montium, quàm alibi, capacissimæ cavernæ reperiuntur.* Pag. 68.

Should any go on to object, That the Waters issuing out of these Caverns, upon their Disruption, would have made but a slender contribution towards raising the mighty Universal Deluge: I answer,

First, They contributed as much to that purpose, as Divine Providence thought fit and necessary.

Secondly, They increased the Waters which ran down the Mountains at the time of the Flood; and so did service in hindring both Men and other Creatures from ascending those Mountains; which might be the *chief* work they were designed to do.

Thirdly, Scripture it self lays the *main* of the Flood upon the *Rain-waters*, ascribing it mostly to them. For so GOD declares, *Gen.* 7. 4. *Yet seven days, and I will cause it to rain upon the Earth, forty days, and forty nights; and every living substance that I have made, will I destroy from of the face*

face of the Earth. Where the great Deluge which was to destroy the then Animal World; is owned as proceeding from the forty days *Rain.* Intimating that the Waters of it, were to rise mainly from *them*; and as for those flowing out of the Fountains of the Deep, they were not to be of equal quantity or use. And indeed had they been so, they would have swelled the Flood to too high a pitch. And therefore though they made but the least part of that fatal Deluge; yet so long as they did what was proper and needful, and what the great GOD intended they should do; that was sufficient.

If, Lastly, it be objected; How could Waters come into these Caverns? I answer; By a very natural and easie way; even the same way that *Springs* do now rise and flow out of Rocks and Mountains. For great Mountains having great Caverns in them, upon the account of their Origination (as being heaved up by the force of that flatuous fermentive moisture (turn'd into vapours) wherewith the Earth at first abounded) how easily would those Caverns be filled with vapours, by the influence of the Sun; and then those vapours condensed into Water, by the coldness of those Caverns? For what were the great Mountainous Caverns, but as it were the Heads of vast *Stills*, as much disposed by Nature to condense Vapours; as the other are by Art. Yea, as cold Water, or wet Cloths, are applied to the Heads of artificial Stills, to help forward their work: So huge quantities of Snow, which outwardly and continually cover the higher parts of some Mountains; might have the like effect on Caverns within. Now these Vapours being thus changed into Waters; the

the Particles of *that* would certainly be too groſs, to ſink down into the Earth again through the little Pores, by which they aſcended or were drawn up out of it. So that unleſs it could find ways, whereby to run forth and diſcharge it ſelf at places in the nature of *Springs*; there it was bound to ſtay till Providence ſhould releaſe it from its cloſe impriſonment; which it did miraculouſly at the time of the Flood, by breaking up the Caverns, or *great Deeps* that contain'd it, and ſuffering a very great deal of it to run out.

So that ſtill the *great Deep Caverns of the Mountains*, may very well paſs for the *Fountains of* Moſes's **Tehom Rabbah**. And that which helps to encourage (not to ſay) and confirm the Notion; is, That *no one* of the ſeveral things, which have been underſtood to be that *great Deep*; can fill up the Character of it ſo fairly, and at the ſame time anſwer the ends and uſes of it, in reſpect of the Deluge, ſo fully; as theſe Caverns. Not the *Open Sea*; for as it could not properly be broke open, as being open already; ſo the Waters of *that* were by no means ſufficient to make ſuch a Flood, as *Noah*'s has been all-a-long reputed. Or in caſe they had been ſufficient; yet being drawn out of the Sea, to drown the Earth, what Waters ſhould have fil'd the Sea again? Or if it ſtood empty, what ſhould have hindred the ſame Waters from running back into it? Not the *Waters in the Bowels of the Earth*: for if they were there in ſuch plenty (as 'tis confeſt there is room enough for them) as to have been able to have made a much greater Flood than *Noah*'s; yet then againſt their nature they muſt have riſen above their Source; and

and being ſo riſen, they muſt have ſtood, ſo long as the Flood laſted, in a miraculous oppoſition to their own nature, inclining them to retire from whence they came. Not the *Supercelestial Waters*: for then the *breaking up of the Fountains of the great Deep*; and the *opening of the Windows of Heaven*; muſt be one and the ſame thing. Whereas by *Moſes* they are very plainly and carefully diſtinguiſht. Not the *incloſed Abyſs*; for then (beſides that the whole *Hypotheſis*, ſo improbable, muſt be allowed) the forty days Rain would have been utterly needleſs. Becauſe then the *falling of the Earth into the Abyſs*, being *the breaking up of the Fountains of the great Deep*; it muſt have fallen in, the very *firſt* day that *Noah* went into the Ark; becauſe on that very day *all the Fountains of the great Deep were broken up*, *Gen*. 7. 11. And if by the Earth's falling into the Abyſs, the World were drowned the firſt day that *Noah* entered the Ark (as of neceſſity it muſt have been, if the Earth were diſſolved and fell that day) to what purpoſe ſhould it after that, rain for forty days together? And whereas it is ſaid, *Gen*. 8. 2. That *the Fountains of the Deep were ſtopped*; the Earth broken down into the Abyſs was never *made up* again, nor the Abyſs it ſelf *covered*; but remains ſtill as *open* as ever: To which Particular Heads, let me add but one more, which has a kind of general Relation to them all. If either the *open Sea*, or the *Waters within the Earth*, or *the Waters above the Heavens*, or the *Abyſs under the Earth*, had been the *great Deep* meant by *Moſes*; none of them had any *true* or proper *Fountains* in them. And ſo what will become of, כל־מעינת, all the Fountains *of the great Deep?*

But:

But now supposing that the *Caverns* in the Mountains were this *great Deep*; how surprizingly do all these things fall in with them? For First, They are called *great Deeps* by the HOLY GHOST (as has been noted) *Psal.* 78. Secondly, They were capable of being cleaved or broke open; as being fast shut up. Thirdly, They were able to afford a competent quantity of Water; even as much as it was necessary they should yield. Fourthly, The Water that came forth of them, could never return into them more. Fifthly, The breaking them up, must be quite another thing, than opening the Windows of Heaven. Sixthly, They might all be broke up the same day that *Noah* took into the Ark. Seventhly, The Rain which fell in the forty days, would still have been as needful as ever. Eighthly, They were stopped again, as strictly and literally, as they were broken up. Lastly, They were as *true* and *distinct* Fountains, as any in the World. So that if they were not the *real* Fountains of the *Mosaic* Tehom Rabbah; one would think they might well have been so.

5. But let us now pass (as it is time we should) to a *Second Ground* upon which we build the *probability* of our *Hypothesis*, above specified; namely, That *the Flood was but fifteen Cubits higher than the highest parts of the surface of the Earth.* And that Ground is this: Supposing *that* to have been the *true height* of the Flood, it will not only be possible, but very easie to find Water enough for it, without recourse to such Inventions; as have been, and justly may be disgustful, not only to nice and squeamish,

squeamish, but to the best and soundest Philosophic Judgments.

For thus, in the *First* place, we need not call in the *Theory*'s assistance; an *Hypothesis* (how ingenious soever in the contrivance and contexture of it) guilty of unjustifiable absurdities.

Nor, *Secondly*, need we fly to a *New Creation of Water*, to gain a sufficient quantity of it. An Expedient that sounds harshly in the Ears of many. And that not only because they are of Opinion, that GOD finisht the work of Creation in the first six days; But because he has expresly declared, That the true and only Causes of the Deluge were these Two; The *breaking up of all the Fountains of the great Deep*, and the *opening of the Windows of Heaven*. To which may be added, That the Creation of so vast a quantity of Water, as should have surmounted the highest Hills; would certainly have inferred, either an enlargement of the whole Universe to receive it; and so a Dislocation, and consequently a disorder of its parts respectively: or else a Penetration of the Dimensions of Bodies; while so much new matter should have sprung into being, more than ever existed; and yet have been confined to the same space of aboad, that was before fill'd up in its whole capacity.

Nor need we, *Thirdly*, to fetch Waters from the *Supercœlestial Regions*. Where, if the Heavens be *Fluid*, how could they have kept from falling down, so long? And if they be *Solid*, how could they possibly have descended at last? For in their descent they must have bored their way through several Orbs as *hard* as Crystal, and how *thick*, we know not. Besides, these Waters must have been lodg'd

S f either

either below the *Stars*, or *above* them. If *below* them, they would have hid them from our sight. The Sun himself cannot be seen through a watry Cloud; how much less the Stars through a watry Ocean? Nor will it help, to say, the Element of Water above is more fine and transparent than the Waters below. For were it as thin as an ordinary Mist, still it would hide the Sun's *Face* from us, though it might transmit his *light*. In case they were plac'd above the Stars, *they* must have been delug'd before the Earth could have been so; as intercepting them in their fall. Nor could they have slid off the Stars again, dropping down to the Earth, unless that were the Center of the Universe, which is hard to prove; yea, most absurd to think.

Nor will it be necessary, in the *Fourth* place, to suppose the *Mass of Air*, or greatest part of it, *was changed into Water*, to make the Deluge. A change which some will by no means admit of, as being not hitherto proved by Experiment. Yet I cannot but own that the best Philosophers have thought it fecible, and also believed it to be actually done. The * *Egyptians* conceived (*Manethus* and *Hecatæus* both attest) ὑετοὺς κατ' ἀέρος τροπὴν ἀποτελεῖσθαι. *That Rains were made by the version of Air.* † *Plato* was of the same Opinion; ἀέρα ξυνιόντα καὶ πυκνούμενον, νέφος καὶ ὁμίχλην. *That Air being thickned and condensed, made Clouds and Mists.* And so was *Philo*. For besides that he affirms * τροπὰς τε καὶ μεταβολὰς παντοίας τρεπόμενος τε καὶ μεταβάλλων, that *it varies and runs through all manner of mutations*: He says expresly in † another Place, πιλουμένου δ' ἀέρος εἰς ὕδωρ, &c. That *Air, being condensed, turns to Water.* And

* *Vid.* Diog. Laert. *Proœm.*

† *In Timæo.*

* *Lib. de Somn.*

† *Lib. de Mund. incorrupt.*

And again, συνιζόντος δ' ὁπότε συνθλίβοιτο εἰς ὕδωρ ἀέρος. That *the Air being condensed may be compressed into Water.* And then brings in *Heraclitus* affirming, τὸν δὲ ἀέρος τελευτὴν, γένεσιν ὕδατος, *the Destruction of Air, to be the generation of Water.* To this also the Lord *Verulam* consents, offering to make it good by sundry * Experiments. Though all of them, I think, come short of Demonstration, or of a clear and satisfactory proof of the *Phænomenon.* (And to name the two greatest Philosophers next.) *Aristotle* asserts this transmutation, in his Book *de Mundo.* And *Des-Cartes* subscribes to it as possible and real. (*a*) *When those Globules move a little slower than ordinary, they may change Water, into Ice; and the Particles of Air, into Water.* And the Famous and Honourable (*b*) Mr. *Boyle* (in his 22. Experiment) leaves it undetermin'd, *whether or no Air be a primigenial Body, that cannot now be generated and turned into Water.* And truly as *Clavius* his Glass of Spring-water (mentioned in that Experiment) Hermetically sealed up for fifty Years past, and reposited in the *Musæum Kercherianum* ; does not prove that Water can't be turn'd into Air, because the Water continu'd there so long without diminution: so neither will M. *Rohault*'s Glass seal'd up the same way full of Air, and kept in a Vessel of Water in a Wine-Cellar three whole Years; argue that Air can't be turn'd into Water, because none of that Air at the three Years end, was found to have suffered such a change; there being not the least drop of Water in the Glass. We only learn from hence, that we have not yet attain'd to the

* *Nat. Hist.* Cent. 1. Exper. 27, 75, 77, 78, *&c.*

(*a*) *Cum isti globuli paulo minus solito agunt, aquam in glaciem mutent, & particulas aeris in aquam.* Princip. *Part.* 4. *Artic.* 48.

(*b*) *Exper. Physico-mechan.*

right Operation, of changing these Elements into one another.

We will grant therefore that by the power of Nature, Air may be turned into Water. Yet neither will *that* take off the whole Difficulty in this Case. For if *most* of the Air incircling the Earth, had been thus changed (and *all* of it could not, because then respiration would have been impossible to Mankind, and the surviving Animals in the Ark) it could not have furnisht Water enough for the Flood; a great deal of Air going to make up a little quantity of Water: (Which the proportion of gravity betwixt Water and Air, of equal bulk, it being (found to be) as of a thousand to one; does sufficiently evince.) But in case it could have yielded Water enough, yet inconveniences would still have remained. Particularly, it would have endangered sucking down the Moon, as the *Theorist* || observes. The changing also of one great Body into another, which after transmutation takes up so much less room than it did before; does either suppose that the whole Frame of the World must sink closer together (which would occasion a strange discomposure in it) to fill up the space *that* Change would make empty: or that in Nature there must be a *Vacuum*. Though (by the way) when our SAVIOUR multiplied Bread upon Earth, *that* need have no such influence on the World, either as to *expansion* or *contraction* of it; as the new *Creation* of Waters above mentioned, or this production of them by *transmutation*, does imply. For besides that the Matter changed was much less in *quantity*; the change might be made in such a Substance, as did take up just the same room in the World before its mutation, as after it.

|| English *Theo*. p. 21.

Nor

Nor need we to apply our ſelves, in the *Fifth* place, to that Hypotheſis which makes this Globe of ours *bi-central*: giving one Center to the *Earth* and another to the *Waters* in it; according to this Figure.

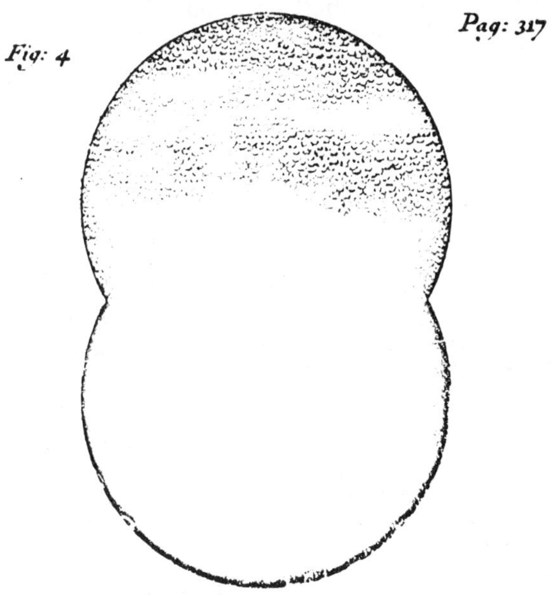

That ſo by ſetting the Waters higher than the Earth, they might the more eaſily overflow it at ſuch a rate, as they have been conceived to do, at the Deluge. But this is a Suppoſition ſo notoriouſly falſe, that to prove it to be ſo, would be a vain work.

Nor

Nor need we trouble our selves in the *Sixth* place about *Sub-terrestrial* Waters. Which (if never so free passages had been opened for them) could no more have flowed up out of the Bowels of the Earth; than Waters can do out of our deepest Wells. Yea, with much more difficulty they must have ascended, in regard they were far deeper in the ground; and also must have boiled up against the weight and pressure of the incumbent Flood, even then when perhaps it was a Mile or two high.

As for Blood flowing out of a Vein (when prickt) in a Man's Head; it is nothing like a Proof that Water may rise and flow above its source. For there is a vital strength and motion forcing it out, and Nature conspires as much to help the Course of that Blood, as she does to hinder this Course of the Waters we speak of. Engines it may be in the heart of the Earth, might be able to send up Waters on to the surface of it; as the Heart in the midst of the Body, sends Blood to its Extremities. But we hear of no Engines made to raise the Flood.

Nor need we, in the *Last* place, to betake our selves to a *Topical* or *Partial Deluge*. A thing which some have done, meerly to avoid the necessity of such a *vast deal* of Water, as they knew not where to have for a *general* Flood, according to the rate of the old *Hypothesis*: or in case they could have had it, knew not how to get rid of it again. Whereas let *fifteen Cubits above the Earth*, be the highest Water-mark of the Flood; and then as the Clouds and Caverns would have yielded Water enough to raise it: so when its work was done, the quantity of this Water would not have been

been ſo exceſſive, but it might eaſily be dried up in that ſpace of time, in which *Moſes* declares it was ſo.

And this is that which in the Second place, gives countenance to our *Hypotheſis*. It makes the Flood to be ſuch, as Nature out of her Store-houſes could very well ſend on to the Earth; and when ſhe had done, as conveniently take it off again. And ſo we are excuſed from running to thoſe Cauſes or Methods, which ſeem unreaſonable to ſome, and unintelligible to others, and unſatisfactory to moſt.

6. A *Third* thing, which gives credit to our Conjecture, and makes it look like truth; is its agreeing ſo handſomely with St. *Peter*'s Deſcription of the Deluge. *The Heavens were of old, and the Earth ſtanding out of the Water and in the Water, whereby the World that then was, being overflowed with Water, periſhed,* 2 *Pet.* 3. 5, 6. How exactly does this ſuite with the *Hypotheſis* propoſed? For according to it, the Earth ſtood partly, *δι' ὕδατος*, *in the Water*; the moſt of it being overflowed; and in ſuch a meaſure, as that the Animal *World* thereby *periſhed.* And yet a great part of the Earth (as much as the upper parts of high Mountains come to) was *ſtanding ἐξ ὕδατος out of the Water*, at the ſame time. Yea, if a *Zeugma* in the words, makes, *συνεστῶσα*, *ſtanding*, relate to, *οὐρανοὶ*, *the Heavens*, as well as to, *γῆ*, *the Earth*; yet our Explication of the Deluge will fall in very fairly with *that* too. Inaſmuch as the Heavens ſtood then *in the Water*, and *out of the Water*, as well as the Earth. For *their* Territories were then invaded in ſome meaſure; the Water riſing, where it incroached leaſt, fifteen;

fifteen; in moſt places, it may be, thirty, forty, or fifty Cubits into them. And therefore ſo high they were ſtanding in the Water; as all above was ſtanding out of the ſame. And which is ſomething more, the Heavens and the Earth will thus be συνεστῶσα, *ſtanding together* out of the Water and in the Water (as ſome will have the word ſignifie there) both of them being in the like circumſtances, at the ſame juncture of time. I will only add under this head, That taking the *Heavens* here mention'd, for the *loweſt Region* of the Air, or for the lower part of that Region; is but conſonant to the Sacred Style.

7. A *Fourth* advantage commending our *Hypotheſis*, is, That it puts the drowned Earth, into a far more habitable condition at the Flood's going off, than otherwiſe it could have been in. That *Noah*'s Flood was Univerſal, is moſt clear from Scripture. *Behold, I, even I do bring a flood of waters upon the earth, to deſtroy* all fleſh *wherein is the breath of life,* from under heaven, and every thing *that is in the earth ſhall die*, Gen. 6. 17. So the ALMIGHTY threatned; and what he threatned he fully made good. *And* all fleſh *died that moved upon the earth, both of fowl, and of cattel, and of beaſt, and of every creeping thing that creepeth upon the earth, and every man. All in whoſe noſtrils was the breath of life, of all that was in the dry land, died. And* every living ſubſtance *was deſtroyed, which was upon the face of the ground, both man and cattel, and the creeping things, and the fowl of heaven; they were deſtroyed from the earth; and Noah* only *remained alive, and they that were with him in the ark*, Gen. 7. 21, 22, 23. And if *all fleſh under heaven*, and *every thing in the earth*, even *every living ſubſtance upon the face of the ground, were deſtroyed and died*;

died; and Noah only remained alive, and the creatures in the ark: hence it will follow, that the *whole Earth* was drowned, or else that Mankind was not generally spread through all the Regions of the same. But that the Earth was generally inhabited before the Deluge, we need not doubt, nor can we well deny. For the Consequence would be, That the Prediluvians begat fewer Children, or lived shorter lives, than the Postdiluvians; which would not be phanciful only, but false. Though truly if some Countries had not been peopled, still they must have been drowned; that so *fowl* and *creeping things*, &c. might be destroyed, according to the Testimony of the HOLY GHOST. Yet admitting this, that the *entire* Earth was overflowed; and that to such an height, that the loftiest Hills (as is commonly believed) had their Tops fifteen Cubits under Water: and what a Case must the Earth have been in, upon drying up of the Flood? What abundance of Mud, Slime, and Filthiness, must every where have covered the surface of it? How thick must it have lain? How close must it have stuck? And how hard would it have been to have clear'd the ground of it? *Attica*, upon this account, (as was observed before) after a far less Flood, was not peopled for the space of three hundred Years.

Nor will the *Theory*'s Explication of the Deluge, help here; unless it be to make things worse. For had the Flood been caused by so strange a *fraction* and *falling in* of the Earth, as that supposeth; this would have added very much to its foulness, and so to its Barrenness for a time (as above remembred) and consequently to its unfitness for immediate habitation.

But now according to the way that we go, the uppermost parts of Mountains could never be drowned; and so never clogg'd neither, or dawb'd over with the

T t filth

filth of the roiled Waters. So that let but the floating Ark, have ſtopped at laſt by the ſide of ſome very large Hill; and the Earth would there have been ready to receive all that came out of it. And that after all its Toſſings, it did reſt *near to*, or in ſome ſenſe *upon* ſuch a tall vaſt Hill (perhaps the biggeſt the Earth has) is rightly believed, as being taught from above. And indeed its doing ſo, ſeems to be no other than a ſignal Providence, and a ſpecial effect of Heavens particular Care. That ſo thoſe few Creatures, which out-lived that grievous πανωλεθρία, or *general deſtruction* that fell ſo heavy on the animate World, might not be deſtitute of fit habitations and ſuſtenance.

And truely that Mankind, upon quitting the Ark, did inhabit Mountains for a conſiderable time; may be gathered (as ſome think) out of the Tenth and Eleventh Chapters of *Geneſis*. For there it appears that they were grown numerous (ſay they) when they left the Hills, and came down to ſettle in the Plains of *Shinar*. But then if they *did* chuſe the Hills for their Seat, and ſtay there ſo long before they removed their Quarters; one reaſon might be the unfitneſs of the lower grounds to entertain them, as affording at firſt no commodious Dwellings. And whereas they would have them to keep on the Hills with deſign to ſecure themſelves from future Floods: ſuch a Deſign would have been utterly vain. For what ſecurity could they expect, by their abode in Mountains, from Floods to come; when the higheſt Mountains were over-top'd no leſs than fifteen Cubits, by one ſo lately paſt?

8. A

8. A *Fifth* Plea which may be taken up in favour of our *Hypothesis*, is its Coherence with Geography. Wherein it seems to be much more plausible than the *old Hypothesis*, or that of the *Theory*. It falls in with it, by a far more natural and justifiable Compliance, than either of *them* do. As for the *Theory*, it flatly denies that there were Hills or Valleys, or Seas, or Islands before the Flood: which Geography hitherto never dreamt of. The old *Hypothesis* also makes the Mountains of *Ararat* or *Armenia*, the highest in the Earth: and this, Geography again cannot down with. And indeed the chief reason why *they* have been reputed the highest, is, because the Ark has always been presumed to rest on the top of them; and in that regard it was requisite they should be the highest. But *our Hypothesis* ties up none to the belief of this neither. Nor indeed does it seem to be worthy of credit, as shall be noted by and by.

CHAP. XVI.

1. Objections *must be* answered. 2. *Our* Exposition *of Scripture not to be made an* Objection *by the* Theorist, *or any that hold with him.* 3. *The* First Objection *from* the Hills being covered, *answered.* 4. *The* Second, *from* the Arks resting upon the Mountains of *Ararat*, *answered.* 5. *The* Third, *from* the appearing of the Tops of the Mountains, upon the decrease of the Waters, *answered.* 6. *The* Fourth, *from* the possibility of Mens being saved from the Flood without the Ark, *answered.* 7. *The* Fifth, *from* the likelihood of other Creatures escaping, *answered.* 8. *The* Sixth, *from* the imaginary excess of Water, *answered.* 9. *The* Seventh, *from* the Raven which *Noah* sent out of the Ark, *answered.* 10. *The* Eighth, *from* danger of Shipwrack which the Ark would have been in. 11. *A* General Answer to farther Objections.

1. WE have seen a *New way* of explaining the Flood proposed : or a *New Hypothesis* concerning it erected. We have seen how it is built; upon what Grounds it stands; and with what reasons and considerations it is supported and establisht. But as things that are new and any whit strange, are commonly received with more than ordinary Notice; so *new Doctrines*, and *strange Hypotheses*, are usually entertain'd with *Disputes* and *Objections*. It will be necessary therefore to look out a little, and to see what Objections are like to meet us in the way

way that we go: and ſo to apply Anſwers to them reſpectively; at leaſt to the chief of them.

2. But firſt, I muſt premiſe, that we have no reaſon to take *this* for an Objection (I mean from the *Theoriſt*, or others who take their meaſures from him) that we expound a Text or two of Scripture ſo as none ever did; and deſerting the common received ſenſe, put an unuſual Gloſs upon them (not to ſay, ἰδίαν ἐπίλυσιν, *a private interpretation.*) This, I ſay, is not to be urged againſt us by the *Theoriſt*, or by *thoſe* that think fit to abide by his *Hypotheſis*. For himſelf exceeds us in the ſame thing. We only take a few ſteps, out of the beaten path of Expoſitors; and that with open and profeſſed diffidence: whereas he has advanced, in as untroden a way, with a great deal of boldneſs.

3. The *Firſt Objection* may be raiſed, from *the Hills being covered.* So we read *Gen.* 7. 19. That *all the high hills under the whole heaven were covered.* And verſe 20. *Fifteen cubits upward did the waters prevail, and the mountains were covered.* Whence it has been concluded, That the Waters of the Flood prevailed to ſuch an height, that they covered the *tops* of the higheſt Hills under Heaven Fifteen Cubits upward. But the Holy Text ſays no ſuch thing. It tells us indeed, That the Waters *prevailed fifteen Cubits upwards*; but this might be meant as to the *Earth* only: upon which, it had told us juſt before, the Waters *prevailed* **greatly**, and *prevailed* **exceedingly**. And truly when they came to be fifteen Cubits upward on the higheſt parts of the ſurface of the Earth; whereby they might be four

four or five times as high above its general *Superficies* (as we have observed) this was really a *great* and *exceeding* prevalence. But where it speaks of the *high hills* and *mountains*, it says no more of them, than, ויכסו, *and they were covered.* And so indeed they were, and fifteen Cubits upward too, that is, on their sides. For the Waters prevailing so high above the surface of the Earth whereon they were founded, the bottoms of them must needs stand up so deep in those Waters. But to affirm the Tops of them did so, is perhaps to make the Comment out-run the Text, *they* being not said to be covered.

And as the *Original* may bear this Interpretation ; so the *Septuagint* seems not to disallow it. For *that* renders the *Hebrew* thus, Πεντεκαίδεκα πήχεις ὑπεράνω ὑψώθη τὸ ὕδωρ, *the water was lifted up fifteen Cubits upwards.* But it does not in the least express, that it was lifted up so many Cubits above the *tops* of the high Hills and Mountains. Nor will the *Vulgar Latin* dissent from it, if rightly understood. It says, *Quindecim cubitis altior fuit aqua super montes quos operuerat. The Water was fifteen Cubits higher upon the Mountains which it had covered.* But then, *altior super montes,* may not signifie, that it was higher *upon the tops* (as was said before) but only *upon* or *about* the *sides* of the Mountains. And so (I remember) when *Q. Curtius* would express Peoples sitting *about* a Table ; he says, They were *super mensam.* And when he would express their sitting *about* a Banquet ; he says, They were *super vinum & epulas.* According to which, *Water fifteen Cubits high* super montes ; may be Water so high *about* the Mountains : and so high indeed it had covered them.

And

And the truth is, the Waters of the Flood never were, nor could be fifteen Cubits above the Tops of the highest Mountains; though we allow the Assertors of the *Old Hypothesis*, to expound the Story of the Flood their own way. To make this out,

We read, *Gen.* 8. 4. That *the ark rested upon the mountains of Ararat*, תנח, says the *Hebrew*; ἐκάθισε, say the LXX. it *sat* there. That is (as the Assertors of the Old *Hypothesis* will have it) * *The bulk of the Ark pierced through the Waters*, and so the bottom of it stood upon the Mountain under it. Nor could it rest or sit there *otherwise*, because the Tops of the Mountains were not as yet above Water, the Flood being at its height. For when was it that the Ark thus rested? Why, *in the Seventh Month, on the seventeenth day of the Month.* And then was the Deluge at the highest. For it is said (*Chap.* 7. 24.) That *the waters prevailed upon the earth an hundred and fifty days.* That is (according to the Assertors of the Old *Hypothesis*) they were increasing, or kept as high as ever for so long time. Which (as the *Jews* used to reckon their Months, making them all τριακονθήμερα, to consist of *thirty days* apiece) will amount to five Months precisely. So that count from the seventeenth day of the Second Month, when the Flood began to come in; to the seventeenth day of the Seventh Month, when the Ark sat upon the Mountains of *Ararat*: and the hundred and fifty days will be expired just. But then if the Ark rested upon those Mountains at *that time*, and in *that manner*, as is said; it is most certain that the *tops* of the *highest* Hills, could never be covered by Water, fifteen Cubits upward. For then if the bottom of the

* *Arcæ moles aquas penetravit.* à Lapide *in Gen.* 8. 5.

the Ark had rested on the Mountains, the whole Body of it must have been quite under Water, and we know not how deep. The reason is, because there are Mountains in the World, very much higher than those of *Ararat*. For by those Mountains, the Assertors of the common *Hypothesis*, generally understand the Mountains of *Armenia*. And the *Vulgar* says expresly, That the Ark rested upon the Mountains of *Armenia*. And the *Septuagint* sometime renders *Ararat*, *Armenia*. Yea, the *Chaldee Paraphrast* uses the word, קרדו, as pointing at the *Cordiaean* Hills. But that there are Mountains much higher than *they*, is evident enough from most considerable Writers. † Sir *Walter Raleigh* declares *that the Mountains of* Ararat, *or any parts of them, are not of equal stature to many other Mountains in the World*. And again, *That the Mountains of* Gordiæi ——— *are the highest of the World, the same is absolutely false*. Nor does he deliver this as his own judgment only, but presently adds, * That *the best Cosmographers, with others that have seen the Mountains of* Armenia, find them far inferior and underset *to divers other Mountains even in that part of the World, and elsewhere*. And then he instanceth in *Athos* as one far surmounting *any Mountain that ever hath been seen in* Armenia: and cites *Castaldus* for it. And to that he adds *Mount* Olympus, *said to be of that height, as neither the Winds, Clouds, or Rain overtop it*. *Solinus*, I confess, says as much, and his authority, I presume, has given credit to the thing. But *Ludovicus * Vives* seems to confute it; and tells us of one, who going up the Mountain to search out the truth of the report, found it to be false. Though when Sir *W. Raleigh* preferred *Olympus*, as

† Hist. of the World, *l.* 1. *c.* 7. §. 10.
† X.
* † XI.
* *Com D. Aug. Civit. l.* 15. *c.* 27.

as to its height, before the *Armenian* Hills; he was certainly in the right. He brings in *Antandrus* alſo, averring *that* for height, to *be of a* far more admiration, *than any in* Armenia. And alſo *the famous Mountains of* Atlas, *ſo high that the eye of no mortal Man can diſcern the top*: for which he quotes *Herodotus*. And laſtly, he concludes the *Pike* of *Teneriff*, to be the higheſt Mountain of the known World. And ſo do others as well as he, whereof *Varennius* is one. Yet ſome again take the *American Andes* to ſurpaſs all. Though *Caucaſus* alone might have done our buſineſs. For as *that* is a part of *Taurus*, as the Mountains of *Ararat* are; ſo it is known to all (who know any thing of that nature) to be much higher than they.

Now the Mountains of *Ararat* being certainly much lower than ſeveral others; here is perfect demonſtration (to the Aſſertors of the general ſtanding *Hypotheſis* of the Flood) that the Waters could not (according to their *Hypotheſis*, and their own Expoſition of the Story of the Flood) cover *all the high Hills* under heaven fifteen Cubits upward, I mean, the tops of them. For then the Ark could not have reſted on the lower ones of *Ararat*, without being plunged under Water, and wholly ſwallowed up; the Flood being at its higheſt pitch when it reſted there.

And ſince it is evident, yea, plainly demonſtrated, that the tops of the higheſt Hills could not be covered according to the tenor of the uſual *Hypotheſis*; it is abſolutely neceſſary, not only in regard of *our Hypotheſis*, but in reſpect to the very Story of the Flood; to interpret *the Mountains being covered*, to ſome other ſenſe than has been put upon it. And that will bring on a like neceſſity of ſetting up a *new Hypotheſis* for explaining the Flood; whether ours may be it or no.

Let us now therefore (as it is neceſſary) enquire after another ſenſe of *the Mountains being covered*. And Firſt, there is a known *Figure* that frequently occurs in the Holy Volume (as might be proved by a large Induction of inſtances) whereby what is true of a thing but in *part*; is notwithſtanding affirmed of the *whole*. And in this ſenſe all the high Hills under Heaven might be ſaid to be covered; becauſe in *part* they were ſo, that is ſo far as the Waters reach'd up the ſides of them. Or,

Secondly, If there be not a *Synecdoche* in the caſe, there may be an *Hyperbole*. The *Mountains* may be ſaid to be *covered*, to raiſe the repreſentation of the Flood, and make it more ſtately, by putting an Air of exceſſive greatneſs into it. So we may obſerve, that there are few very grand and remarkable things in Scripture; but the Mountains or Hills are brought in, to bear a part in their Character: to adorn and ſignalize, or ſet off their magnificence or exceſs. Thus a great *ſteadineſs* is expreſs'd by the *ſtability* of *Mountains*, *Pſal.* 125. 1. *They that truſt in the* LORD, *ſhall be as* mount *Zion, which cannot be removed.* A great *ſafeguard* or *protection*; by an *incloſure* or *incompaſsment* with *Mountains*: as in the next Verſe. *The* mountains *are round about Jeruſalem; ſo the* LORD *is round about his people.* A great *deſtruction*; by the *trembling of Mountains* and *removing of Hills*, Jer. 4. 24. *I beheld the* Mountains, *and lo they trembled, and all the* Hills *moved lightly.* So a great *fear* is expreſſed by Mens calling out *to* Mountains *to fall on them*,

them, *and to* hills *to cover them*, S. Luk. 23. 30. A great *change*; by the passing away or *disappearing* of mountains, *Rev.* 16. 20. A great *victory*; by *threshing the* mountains, and *making the* hills *as chaff*, *Isai.* 41. 15. A great *joy*; by *the singing of* mountains, *Isai.* 44. 23. A great *slaughter*; by Blood reaching to the *Mountains*. So GOD threatens to *water the land of* Egypt *with blood*, *even to the* mountains, *Ezek.* 32. 6. And that the mountains *shall be* melted *with the blood* of nations, *Isai.* 34. 3. As if Blood were not only to swim *about* the Mountains, and to run *over* the *tops* of them, as *Noah*'s Waters (are presum'd to have done) but even to dissolve them and wash them quite down. Well might *Moses* hyperbolize as he did, in describing the Deluge of Water; when the Prophet thus exceeds him in foretelling an Inundation of Blood.

By no means, may some object: and you have hinted the *reason* of it. Even because what *Isaiah* spake was in way of *Prediction*; and such *Hyperboles* though they be common in *Prophecies*, are not used in *History*. I answer, Such *Hyperbolical* Schemes of Speech, are used in *Historical*, as well as in *Prophetical* matters. Thus the *Psalmist* referring to the majestic or great solemnity at the Promulgation of the Law; says, *the* hills *melted like wax*, Psal. 97. 5. And relating some circumstances of *Israels* passing out of *Egypt*, he says the mountains *skipped like rams*, *and the little* hills *like lambs*, Psal. 114. 4. And if against this it be objected, that the *Psalms* are *Poetical*, and so these may be flights of Phancy, allowable to *Poets* only; (though to inspired ones, as well as to other:) I answer, The like occurs in other Books of

Scripture. *Isaiah*, for instance, reflecting upon great and terrible things that GOD had done for his People; sets them forth by this Expression, *The* mountains *flowed down at thy presence*. Isai. 64. 3. And *Habakkuk* commemorating GOD's miraculous proceedings in bringing the *Israelites* into *Canaan*; says, *the everlasting* mountains *were scattered, the perpetual* hills *did bow*, Hab. 3. 6. And at the *tenth* Verse, *the* mountains *saw thee, and they trembled*.

So that when *Moses* described the Deluge, in so superlative or transcendent a strain, as by its *covering the* mountains, *and all the* high hills *under heaven*; it might be but to ingrandize or amplifie the thing. It might be but an high flying ornamental *Hyperbole*, used to grace and greaten the Flood in his Description of it, and to render it the more stately. Or,

Lastly, We must know; that to *cover* a thing in Holy Style, is not always to *surmount* and *overwhelm* it: but very frequently to surround it only, or to be *about* it in *great plenty* or abundance. For so the HOLY GHOST does commonly express the *copiousness* of one thing by its *covering* another. Thus *precious stones*, are said to be a *covering* to the *Tyrians*, Ezek. 28. 13. because they wore them in great plenty about them. And the *Jews* are said to cover *the altar of the LORD with tears and weeping, and with crying out*, Mal. 2. 13. because they shed their Tears, and uttered their complaints very freely and plentifully thereabouts. So (in the same sense) some are said to be *covered* with shame; and others, to be *covered* with confusion; and others, to be *covered* with violence. Whence it is evident, that it is a Phrase whereby

is expressed the *plenty* or exuberance of one thing above another. And so the *Mountains* and *high Hills* being *covered* with Waters, will signifie no more, than that they were surrounded with vast quantities of them. But a more adequate and evictive instance of this, *Moses* himself (the fittest Man that could be in the case) has given us. *Who*, setting down the Story of the *Quails*, Exod. 16. 13. says, That *at even the quail came up and* **covered** *the camp.* And how did these Quails *cover* the Camp? He informs us, *Numb.* 11. 13. they *fell by the camp as it were a days journey on this side, and a days journey on that side,* **round about** *the camp.* But then as the Camps being *covered* with the Quails, was no more (in *Moses*'s language) than its being *surrounded* with a *multitude* of them; so the Mountains being *covered* with Waters, was no more than their being *surrounded* with *great plenty* of them. And, ויכסו, which signifies, *covered*, Gen. 7. 19, 20. and, ותכס, that signifies, *covered*, here, *Exod.* 16. 13. do both spring from, כסה, and are but one and the same word. As if by using the same word in both places, he would intimate, that he meant but the same thing in both Stories.

4. A *Second Objection*, may be *the arks resting upon the mountains of Ararat*, Gen. 8. 4. For that implys that the Waters of the Flood did certainly swell up above the tops of those Mountains; else how could the Ark have been carried up thither, and have rested there? I answer,

First,

First, That, על, which is there rendred, *upon*; does sometimes in Scripture, signifie, *by*. As על־פלגי־מים, by *the rivers of water*, Psal. 1. 3. על־המחנה, by *the camp*, Numb. 11. 13. And so here, על־הרי אררט, may signifie no more than, by *the mountains of* Ararat.

Secondly, If we yield the Ark to have rested *upon* the Mountains of *Ararat*; yet then it might rest somewhere upon the *foot* or lowest part of those Mountains: for it is no where said to rest upon the *top* of them. And so this passage in the *Divine* Story, will not infer the least necessity of that vast height of the Flood, which it has usually been set at.

And then as to *Civil* Story, which tells how the Ark rested on the top of those Mountains; we take leave to observe, that it is all-a-long chargeable either with *incertainty*, or with *incongruity*. It is still either *Doubtful*, as to the *Thing*; or *inconsistent* with *it self*. Thus, for example, * *Josephus* gives account out of *Berosus*, τὸ πλοῖον, &c. that *a part of the Ship* (the Ark) *is in* Armenia, on the Mountain of the *Cordiæans*. But then this is ushered in with a, λέγεται, so *it is said*; which makes the thing *doubtful*. And then, πρὸς τῷ ὄρει, may as well be rendred, near *the Mountain*, as *on* it. The same *Josephus*, in the same Chapter also, thus certifies out of *Nicolaus Damascenus*. Ἔστιν ὑπὲρ τὴν Μινυάδα, &c. *Above* Minias *there is a great Mountain in* Armenia *called* Baris, *on which many that fled thither were saved in time of the Flood: and that a certain Man brought in an Ark, arrived at the top of the Mountain, and that the reliques of the Timber were kept there a long time.* But then this is delivered *incertainly*

* Antiq. l. 1. c. 4.

incertainly again, with a, λέγεται ἔχει, so *it is reported.* And for a matter to be *reported*, is one thing; and really to *be so*, is another. And indeed this report agrees not with truth. For it says, εἰς ὃ πολλοὺς συμφυγόντας ἐπὶ τοῦ κατακλυσμοῦ διασωθῆναι: *that* many flying *to this Mountain were saved.* Whereas Scripture on the contrary assures us that *but few* were saved: and that not *one* was saved by *flight*, but all by the Ark alone. And then it is *inconsistent* with it self too. For how could the Ark drive ἐπὶ τῆς ἀκρώρειαν, up the very *tip-top* of a Mountain if there were no Water upon it; but so much dry ground, as that many might be saved there, living together, not only Days, and Weeks; but several Months, one after another.

Eusebius likewise and *Cyril* do both recite out of *Abydenus* the *Assyrian*, how τὸ πλοῖον ἐν Ἀρμενίῃ, *the Ship* (Ark) *in* Armenia, *did out of its Wood afford Amulets to them that dwelt thereabouts.* But then the other circumstances of the account are strangely odd and fabulous. Namely, *That Saturn who reigned at that time, forewarned* Sisithrus (Noah) *that there should fall abundance of Rain on the fifteenth of* Desius; *and commanded him to hide what learned writings he had, in* Heliopolis. *Which,* Sisithrus *having done, he sailed directly into* Armenia, *and there quickly found what the GOD had told him, to be true. But then on the third day after the Tempest, sending out Birds to try whether they could see any Land that was not covered with Sea; they returned again, as not finding any place where they could rest. After them he sent forth others; and when he had sent the third time, the Gods took him away from amongst Men.* Where the absurdities and incongruities

incongruities of the Story (if brought home to the truth of things) are so many, and gross, and obvious; that time would be perfectly lost, should I spend any in noting them.

I pass therefore to *Benjamin* the *Jew*, from whom I borrow the last citation of this nature. He says in his *Itinerary*, that the Ark of *Noah* rested upon the Hills of *Ararat*; and that one *Omar*, of the Materials of it, built a *Mahometan* Synagogue. But then he adds that the Prince took it down, *è Cacumine duorum montium, from the top of* two *Mountains*. And that the Ark should be divided into *two* parcels, and the remains of it lodg'd upon the tops of *two* Hills at once; is a passage that gives but small Credit to the Traveller's Report; but is enough, methinks, as to this Particular, to call his fidelity into question.

Notwithstanding therefore what we meet with in History concerning it, we may lawfully conclude, that the Ark might not rest upon the *top* of the *Armenian* Mountains. Only one or two Writers of note, mistaking *Moses*, it may be, at first; and telling the World with confidence, That the Ark rested on the *top* of these Mountains (when he might mean no more than that it rested *by* them, or on some *low Ridge* of them) others might follow *them*, and others *them* again; and so *all* might run on in a Track of error, as smoothly as if they had been in the way of truth. Thus, when St. *Chrysostom*, *Epiphanius*, *Isidore*, and others, tell that the Ark rested upon the *top* of these Mountains; and that certain, λείψανα, *remnants* of it were to be seen there in *their* days: they were probably over-rul'd by *History* or *Hearsay*; and so easily mis-led mis-led by such as went before them. And indeed that the thing was utterly false, we have great reason to conclude; when if it were true, it must either impeach *Scripture*, which (in the sense of all Men hitherto) taught all-a-long that the Flood was fifteen Cubits above the tops of the highest Mountains; and that in the height of this Flood the Ark rested on the top of *Ararat*: or else clash with *Geography*, which never allowed the Hills of *Ararat* to be (by a great deal) the highest: or else *sink* the Ark quite under Water, to make it *rest* upon those Hills.

5. A *Third Objection* may be formed from *the appearing of the tops of the Mountains upon the decrease of the Waters.* So it is recorded, *Gen.* 8. 5. That *the waters decreased continually until the tenth month, and on the first day of the month were the tops of the mountains seen.* Now if the Mountains had not been quite under Water, and so invisible for the time they were overwhelmed; how could they be said to become visible again, or to be seen upon the Floods going off?

In answer to this, we may consider, First, That by the tops of the Mountains, in Scripture, are not always meant the *higher*, but sometimes the *inferior* parts of them. Thus it is prophesied, *Amos* 1. 2. That *the top of Carmel shall wither.* Where by *top*, the *sides* or *lower parts* of that Hill may be intended chiefly. For the withering of the meer top of it only, would not ('tis like) have either caused or signified such a scarcity of feed, as should have occasioned such affliction to Shepherds, as is there foretold; the principal part of an Hill for Pasture, being usually towards the bottom of it. So *Exod.* 19. 20. it is said, That *Moses went up to the* top *of mount Sinai.* But that he did not go up to the *very top* of that Mount we have great cause to believe, for Two reasons. *First*, Because *the LORD descended upon it in fire*, ver. 18. in such a fire as was not only *real*, but *raging*; for it made the Mountain *smoak as a furnace.* Yea, it is said, *Exod.* 24. 17. to be *like devouring fire on the top of the mount.* And so *devouring* was it, that it seiz'd most terribly upon the Mountain; insomuch that *it* is said to have *burnt with fire unto the midst of heaven*, *Deut.* 4. 11. But how then could *Moses* go up to the *top* of this Mountain? Nor, *Secondly*, could he well do it, by reason of its *height*, and the great *difficulty* of its *ascent.* For *Josephus* assures us, That *it is the highest Hill beyond comparison of all that Country, and long of its strange height, and its steep inaccessible craggy Rocks, is not only unfrequented by Men, but not to be lookt up to, it puts the eye to such pain.* And yet if *Moses* did not go up to the *top* of this Hill in *strictness*, we know not how much below it he might present himself. And in case he stood on any lower ridge or part of that Mount; it is clear that by the *tops* of high Hills in Scripture, may be meant but the *lower parts* of the same. And therefore where we read ומראש הררי, *Deut.* 33. 15. *from the top of the mountains*; the *Arabic* reads it, *from the roots* of them. And so by the tops of the Mountains being seen upon the drying up of the Flood; will be meant no more, than that some lower parts of them, not far from the bottoms, were made bare and expos'd to view again,

again, which before were hidden under Water. Or,

Secondly, By the tops of the Mountains, ſaid to be ſeen on the firſt day of the tenth Month; may be meant, but the tops of ſome *lower* Mountains, which were quite overwhelm'd with Water, by its aſcending fifteen Cubits upward upon the higheſt parts of the plain of the Earth. If theſe *two* Conſiderations will not ſatisfie; we muſt carry on the Enquiry a little farther, and ſeek for a *Third*. And truly ſome one or other muſt needs be found out. For certain it is, that the tops of the higheſt Mountains could not be ſaid to be ſeen, by reaſon of the Waters ſinking down below them; becauſe, as we have ſufficiently proved, they could not poſſibly be above them. That is, according to the common meaſures Men have taken of the Flood, and the uſual ſenſe they have put upon the ſacred Story of it.

Thirdly, Therefore (in way of anſwer to the Objection) we conſider; that the tops of the Mountains may be ſaid to be ſeen, at the time mentioned, upon account of their *emergency out of darkneſs, not out of the Waters.* Nor let it ſeem ſtrange, that at the time of the Flood, there ſhould be *darkneſs* over the whole face of the Earth. For then there was a ſolution of the continuity of the *Atmoſphere*: all the vapours almoſt contained in the περιοχῇ, or comprehenſion of it, turning into Clouds, and reſolving a great pace into Rains. And as it is but *reaſonable*, to think it was dark then (conſidering the ſtate of the Atmoſphære) ſo it was very *requiſite* it ſhould be ſo. For when the Rains began to fall, and that at ſuch a rate, as to threaten in good earnest to make that Deluge which *Noah* had foretold: this must needs startle and alarm Men dreadfully. Then, had there been light in the World, in any good degree; what could have been expected, but that People who dwelt nearest to that place where the Ark stood; should have run directly to it, and rudely assaulting and invading it, have turned out *Noah*, his Friends, and all Creatures; and have taken immediate possession of it themselves, as the only probable means of their own preservation. And therefore that the Earth was then wrapt up in nightsome darkness (it being not only *likely* in respect of *Nature*, but *necessary* in point of *Providence*) we need not fear to conclude.

And as it was dark all the time that the Flood was coming in and waxing; so the Air might well be very foggy and misty during the continuance and decrease of the same. For the Atmosphære being put into so great a disorder (and even dissolution) as it was; it could not quickly resettle into its wonted clearness.

And then we must heedfully attend to that account of the Floods abatement and drying up, which the HOLY GHOST has given us. *The waters returned from off the Earth continually*, says he, *Gen.* 8. 3. Where, the word, ישבו, *returned*, does often signifie in Scripture, the returning of a thing into its *Principles*. So *Psal.* 90. 3. שובו בני־אדם, return *ye sons of men*. As much as to say, be resolved into Dust and Spirit, the primigenial parts, or constituent principles of your Nature. And *Gen.* 3. 19. it is used in the like sense; אל עפר תשוב, *to dust shalt thou* return. And *Psal.* 146. 4. ישב לאדמתו, *he* *he* returneth *to his Earth.* According to which, where it is said, That *the Waters* returned *from off the Earth continually*; we are to understand their continual *version* or return into that *Principle* out of which they were made; namely, *Vapours.* And the same is to be understood concerning them, where it is said, *Gen.* 8. 5. That *the waters decreased continually.* הוו הלוך וחסור, *They were* going *and decreasing.* And so the Expression does not denote a violent *motion* or *agitation* of those Waters (as hath commonly been thought) so much as a constant *wasting* or *diminution* of them, by *going quite away.* And indeed, הלך, signifies, it *went away*: and as * *Schindler* notes, is spoken *de rebus evanescentibus, of things that are vanishing.* Yea, the learned *Lexicographer* brings in this very Passage, as one instance of that its signification. Which farther insinuates, That when the Waters of the Flood decreased, it was done by their *vanishing* or *going away* into their first natural *Principle*: by their *returning* or being *converted* into *Vapours.* Now this being done at a *great rate,* or *very fast* (as we may gather from so much Water being dried up in so short a time; and from the miraculous Wind, *Gen.* 8. 1. sent on purpose to hasten the work, by helping forward the attenuation of the liquid Element) it must (in likelihood) overcast and bemist the Air; and so conspissate and obscure it, as to render things invisible at a little distance from the beholder's Eye. Whence it will follow, That when the tops of the Mountains were seen, this might come to pass, not by the Waters sinking below those tops (whither they never ascended) but by the clearing up of the Sky, and the wearing off

* *Ad vocabulum in Lexic.*

off of its unusual thickness and fogginess. And yet this their *visibility* or *new appearance* might properly be ascrib'd to the decrease of the Waters too: inasmuch as till they were so diminisht, as not to afford Vapours enough, to thicken and darken the Air any longer, at the rate they had done; the Mountains tops could not be seen.

Should it here be objected, That according to this way of explaining their appearance, they could not have been seen so soon as in the tenth Month; because the Waters were then upon the Earth in great abundance: that Objection might be thus taken off. Though there were *waters* * *upon the face of the whole earth* then; yea, and † forty days after that (which was the reason why *Noah*'s Dove could find no rest) yet these Waters were so far exhaled, drawn so low, and grown so gross and muddy; that now they did not *return* or *go away* into vapours, half so fast as before. The Atmosphære also was now come pretty well to its old consistency again; and so the attractive power of the Sun was much damped and weakened, and he did not draw vapours so briskly and plentifully as he had done. And yet the lower Regions of the Air might be very thick and foggy still; so that the Mountains might not be seen by looking right on, but rather by looking upward. And so the highest parts of the Mountains, that by thrusting up aloft did intercept the lightsomeness of the glimmering Skie, and terminate the eye-sight; might by that means be discerned. And therefore indeed only the *tops* of them were said to be seen.

* Gen. 8. 9.
† Ver. 6.

Nor

Nor let it be thought a meer phancy, a whimsical groundless Figment of ours, that the Waters of the Deluge did decrease in this manner. I mean by *going* or returning into Vapours, and that at such a rate as to fill the Air, for a time, with constant Mists, and make it very caliginous and dark. This is so far from being an empty fiction or conceit; that I may venture to say, It was a necessary *Phænomenon.* For when the Earth was so generally drown'd, the Water being of a smooth Superficies, if the *Air* had been clear, yea, if it had not been more than ordinarily thick, it would certainly have been most exceeding cold. Even as cold as it is now in its middle Region, where Icy Meteors are continually floating. So that in the Natural Course of things, the Waters of the Flood would presently have been frozen extreamly hard. And if we can suppose they should ever have been melted again (as by the force of meer Nature they hardly could) yet they could not have been so in that space of time, wherein the Deluge went off, and the Earth became dry.

And that a vehement Frost would have seiz'd the Waters of the Flood, as soon as they were come down (if the Air had not been strangely thick) is but reasonable to conclude upon *this* account. Because the Atmosphære was never so exhausted of Vapours; and so never so thin; and so never so sharp and terribly cold, since the World began; as it was at that time.

And then lastly, that the closeness and thickness of the Air was such, as to darken and benight the whole Earth at once; may fairly be inferred from *Gen.* 8. *ult.* For there GOD promiseth that while

while the Earth remaineth, there shall be *day*, as will as *seed-time*, and *harvest*. Implying, That during the Flood, there was as perfect an intermission of *day* upon Earth, as there was of *seed-time*, and *harvest*.

6. A *Fourth Objection* may be framed from the *Possibility and easiness of Mens escaping the Flood.* For if the Waters prevailed but fifteen Cubits upwards upon the Plain of the Earth; and the tops of the spacious aspiring Mountains stood bare (excepting a little of the lower parts of them) all the time of the Deluge: how easily might Men have run up those Mountains, and so have been saved from the violence of the Waters? and then what need of an Ark to preserve them.

To this it may be answered. For People to ascend these high Mountains, when the Flood was coming in; could be no such easie matter. For at what rate soever the Rains descended in other places; it is not to be doubted but they fell in great abundance about the lofty Mountains. For the pitchy, swollen, loaden Clouds, which then hung every where bagging in the Air; driving and crouding, and squeezing against those Mountains, could not but empty themselves there (like full Spunges when pressed or nipped) in prodigious Showres, that would have run down in furious and mighty Torrents. Yea, 'tis more than probable, that these squeezed Clouds, would not only have discharged themselves in immoderate Showres thereabouts; but in kind of *Ecnephiæ*, or *Exhydriæ* (such as sometimes fall in the *Pacific* Ocean) very terrible Tempests; wherein Rain pours down as it were out of Spouts

Spouts or Buckets, and falls in whole Sheets of Water at once. So that the ſides of the Hills would have been full of Cataracts, and the Waters would have come roaring and gulling down them ſo forcibly, that no living Creatures would have been able to ſtand, much leſs to climb up againſt them. And then the higher ſort of Mountains, as the *Alpes*, and the like, being covered with huge quantities of Snow; that would have melted a great pace too, and contributed to the dreadful Torrents we ſpeak of. And then the Waters of the *Great Deep*, being no other (as we ſuppoſe) than ſuch as flowed out of the *Caverns* of high Rocks and Mountains, when the power of Heaven had broke them up: theſe alſo would have augmented the mighty Defluxions, and made them more violent and irreſiſtable. And this was one main end of GOD's breaking up thoſe Fountains; even to increaſe the Downfals of Water off the Mountains, and to make them ſo copious and fierce, as that Men might not be able to aſcend the Mountains. And truly for them to have fled to the Mountains to be ſaved from the Flood, down which ſuch impetuous Streams came rolling and roaring in moſt hedious ſort: would have been like plunging themſelves into the Sea, to prevent drowning.

And truly if any Houſes, Towns, or Cities, ſtood ſo high upon Mountains, as to be above the Water-mark of the Flood: yet the aforeſaid Downfals of Water, would have ruined them all. Or if any could have ſupported themſelves by their great ſtrength, the Inhabitants would ſtill have been drowned in them. Which might be one main Reaſon, why GOD appointed *Noah* to build an *Ark* (and not an *Houſe*, or a *Caſtle* upon any high Mountains) to ſave himſelf, and ſuch other Creatures as were to be preſerved.

7. A *Fifth Objection* may be drawn from *the likelihood of ſome other Animals eſcaping the Flood.* That is, ſuch as lived *within* the Earth, in the upper and undrowned parts of the Mountains. For however they could not get up on the Hills, or if they had been upon them, could not have harboured there; but muſt have been waſhed down into the common Gulf that ſwallow'd all : yet having their aboad *under ground*, and perhaps a *good depth* under it too; they might be ſecure in their ſubterraneous Dwellings. For though the Waters fell in great plenty, and with as great violence; yet ſhedding off the Mountains apace. and haſting downward ſwiftly; they could not ſoak ſo far into the Earth, as to incommode, much leſs deſtroy the Creatures there lodg'd, and ſo well intrencht and fortified againſt them. The Conſequence would be no leſs, than that *Moſes* muſt faulter in what he relates; That *every living ſubſtance was deſtroyed*, Gen. 7. 23.

I anſwer. Where the Hiſtorian tells *this*, that every living ſubſtance was deſtroyed; he immediately puts a reſtriction or plain limitation upon it; adding אשר על פני האדמה, *which was upon the face of the ground.* So that if any creatures were ſo deep *under ground*, as to continue alive and ſafe, notwithſtanding the Deluge: this would be no contradiction or repugnancy to the Inſpired Writer. For ſtill *every living ſubſtance* might be deſtroyed which

which was *upon the face* of the Ground: and that was as much as he affirmed.

Least that Anſwer ſhould not ſatisfie, let me put in another. The Waters falling ſo plentifully and violently on the Mountains; where they could not ſoak in, and drown the Creatures earthed in them; by their continued beating and running upon the Ground for forty days together, they did either ſo ſettle it, that it ſqueezed them to death: or elſe ſo ſtop up the pores of it, that they were ſmothered.

8. A *Sixth Objection* may be taken from *the Quantity of Waters as like to exceed very much* (ſome may imagine) even ſo as to ſurmount our *ſuppoſed limit*. For they that iſſued from the Fountains of the Great Deeps, joined with thoſe that fell in the forty days; muſt needs have raiſed a Flood much higher than fifteen Cubits above the Plain of the Earth.

But the anſwer ſays, No. For beſides the huge deal of Water which the Earth drank up (eſpecially in its ſandy Regions) before its thirſt could be quenched; and the vaſt deal that ſank into its inviſible hollowneſſes, before they could be filled; and the abundance that was abſorpt by its numberleſs pits and capacious valleys, before they could be repleniſht, and the Water brought to a level: And beſides how much it then took up, to raiſe the Flood *one* Cubit around the Globe, as well upon the Sea, as dry Land; and how much more to raiſe a *ſecond* Cubit, than the firſt (the higher circumference being ſtill the larger;) and how much more to raiſe a *third* Cubit, than the Second; and

 ſo

so on till the fifteen Cubits were full: Besides all this, I say, the Rains by which the Deluge was *chiefly* caused, might not descend at any *extraordinary rate* of violence. For however about the Mountains, they might be monstrous and intolerable; yet every where else they might be quite otherwise: and the immensity and destructiveness of the Waters they raised, may be imputed to the generality and duration of them, rather than to their excessive greatness. We are told indeed, *Gen.* 7. 11. That the *cataracts* or *windows of heaven were opened.* Yet that might betoken nothing extraordinary in the Rains, save their *continuance.* For *Mal.* 3. 10. GOD promiseth his People (as a signal mercy) to *open* (ארבת) *the cataracts* or *windows of heaven* for them. And what does the Expression there import? Why, no more than that he would send such moderate Rains, as should make their grounds fruitful. So says *Lyra*; GOD opened the Cataracts of Heaven, * *by giving rains and dews convenient to make the ground fruitful.* And if *the opening of the cataracts of heaven*, implys but an ordinary descent, or moderate downfal of gentle fructifying Rains and Dews: then notwithstanding these Cataracts were opened at the Flood, the Rains might then in most places distill, with a wonted gentleness and moderation. Which granted, there would be no danger of their swelling the Flood above that height to which our Supposition limits it. And though according to *Marsennus*'s account, forty days Rain might raise the Waters an hundred and fifty Feet: yet who can tell whether the Rains fell so fast in those forty days; as they did at the time, and place, when, and

* *Dando pluvias & rores ad fecunditatem terræ convenientes.* In loc.

and where, he made his Experiment and Calculation? Others I am ſure are of the mind (and *Oſiander* for one) that they were only ſufficient to ſet the Ark afloat. And they quote that Paſſage for it, *Gen.* 7. 17. *The flood was forty days upon the Earth: and the waters increaſed, and bare up the Ark, and it was lifted up above the Earth.*

9. A *Seventh Objection* may be made from *the Raven which* Noah *ſent out of the Ark*, Gen. 8. 7. It is there ſaid, That *that* Raven *went forth to and fro until the waters were dried up from off the earth.* Whence ſome conclude, That he was forced to return into the Ark again and again, ſtill as he went out, becauſe by reaſon of the Waters, there was no convenient place of abode for him abroad. And conſequently they infer, That the Waters which were ſo high *then*, could not but cover the tops of the Mountains, when they were at their full height. To this it might be anſwered,

Firſt, That if the Raven *did return*, this does not argue that the Waters were then at ſuch a *mighty height* (and ſo that they had been higher than the *loftieſt* Hills) becauſe it is ſaid, That he *went to and fro* (that is, to and from the Ark, as our Objectors would have it) *until the waters were dried up.* So that his *returning* was not occaſioned by the exceſs of Waters, not ſuffering him to remain at large; nor does it prove them to have been ſo exceſſive as they would make them. For even when they were *abated*, and ſo abated that the tops of the Mountains were ſeen (ver. 5.) where he might have had both *reſt* and *prey*; ſtill (according to the *Hebrew Phraſe*) he was *going and returning* from and to the Ark.

Ark. Yea, he continued to do thus *all-a-long*, even *untill the waters were dried up from off the earth.* Which makes it plain, that as the excess of the Waters could not be the cause of his returning to the Ark; so his continual returning could not argue the Waters to be so excessive: inasmuch as he never ceased returning, till the Waters were quite dried up. But,

Secondly, I answer. The Raven in likelihood returned not at all. And therefore the *Vulgar* is positive in the case; *egrediebatur & non revertebatur*; *he went out, and did not return.* And so is the *Septuagint*; ἐξελθὼν οὐκ ἀνέστρεψεν. And *Bochart* says, That if the *Negation* be taken out of the *Original* Text, there will be no sense in it. And therefore he thinks that יצוא ושׁיב ought to be, ולוא ישוב. putting the *Future* Tense for the *Præterperfect*. And then the Raven for certain did never return to *Noah*. And the *Arabian* Proverb intimates as much, אבטא מן עראב נוח. he *stays as long as* Noah's *Crow*. To which the *Latin* one is near akin, *Corvus nuncius*: Or, *Corvum misimus*. So that the Objection *against* us, will at last be a piece of an Argument *for* us. So far, that is, as the Raven's not coming home again, after he was sent out; shows the Waters were low: and that he had Food enough to live upon, and Room enough to fly up and down in from place to place; which might be that *going and returning of* his, mentioned, *ver.* 7.

Indeed the Dove which was sent out after, *found no rest for the sole of her foot, and* therefore *she returned to Noah into the ark*, verse 9. And no wonder. For though the Waters were much abated, yet still they were *on the face of the whole earth*, covering its Superficies in most places. And the Dove being a more

more nice and tender Creature than the Crow; might want proper Food and a warm Rooſt, and for the ſake of theſe, be glad to fly back to the Ark where it had found both. And therefore the ſecond time that it was ſent forth, it returned not till the *evening*; that is, till the coolneſs of the approaching night, made it ſenſible of the want of a convenient Lodging. And for the ſame reaſons (eſpecially it being a tamiſh Bird) it might perhaps have come back to *Noah*, when he ſent it out laſt: only the Earth and Air being now grown more dry, and warm, and pleaſant; probably it was tempted to fly ſo far from the Ark, as not to be able to find the way to it again. Yet its not returning might be *really* to *Noah*, what he took it to be; a ſign that the Waters were dried up.

10. An *Eighth Objection* may be *the Danger the Ark would have been in, of being ſtav'd or wrackt.* For if during the Flood, the tops of the great Hills had been all above Water; how eaſily might the Ark have run aground, and have been broken and ſhattered all to pieces?

It may be anſwered thus, The great Deluge from the Beginning to the End of it, was in great meaſure a *miraculous* work. Yea, even where GOD was pleaſed to make *Nature* his Inſtrument; He took her, as I may ſay, into his own hand, and wielded her by his own Omnipotent Arm; and ſo inabled her to do, what in her own way, and by her own ſtrength, ſhe could never have effected. Look into the inſpired Story, and what a great deal of miracle ſhall we ſee, in the very Præluſories or preparatives to that mighty Inundation?

Thus,

Thus, as GOD preacquainted the Patriarch *Noah*, with his design of bringing it in; so he ordered him to build an Ark against it came, to save himself and his Family, from that fearful ruine which was to attend it. He directed him of what Timber to make it, and of what Dimensions; how to frame it without, and to fashion it within: and the whole Vessel seems to have been all of his wise contriving. Such Creatures also as were to be kept alive for future propagation, he appointed *Noah* to admit into this Ark; inclining them at the same time, to come in their several species, and offer themselves to him. For as the Father says; * Noah *did not catch them and put them in, but when they came and went in, he suffered them* to do so. And thus much he will have signified, *Gen.* 6. 20. *Two of every sort shall* come *unto thee. Non scilicet hominis actu, sed Dei nutu. Not by the diligence of man, that is to say, but by the disposition of* GOD. And as he injoined *Noah* to receive these Animals into the Ark, and harbour them there; so likewise to provide sustenance for them, instructing him as to the quality and quantity of the same. So says the same Father. *(a) What wonder, if that wise and righteous man who also was divinely taught what was agreeable to every creature; did procure and lay up sutable nourishment to every kind?* And to the end he might have all in a due readiness against the time, GOD gave him a *weeks* notice, just before the irruption of the fatal Waters, *Gen.* 7. 4. And lastly, when the good Man and his Relatives entred the Ark (whose *Cargo* was such, as no single Ship, nor the mightiest Fleet could ever boast of)

* *Non ea* Noe *capta intromittebat, sed venientia & intrantia permittebat.* Aug. *de Civit. l.* 15. *c.* 27.

(a) *Quid mirum si vir ille sapiens & justus, etiam divinitus admonitus, quid cuique congrueret; aptam cuique generi alimoniam praeparavit & recondidit?* Ibid.

of, though the Sea it navigated was as wonderful, as its Lading) the LORD himſelf is ſaid to *ſhut them in*, Gen. 7. 16. That is, by the Miniſtry of his Holy Angels.

And when the ALMIGHTY was thus *miraculouſly* ingag'd in ordering the *Preparatives* to the Flood; we may be ſure it was no leſs concern'd in bringing in the Flood it ſelf. And therefore GOD openly proclaims it to be his own Fact, and challenges and appropriates it to himſelf alone, as peculiarly belonging to his Providential Efficience, *Gen.* 6. 17. and 7. 4. And St. *Peter* expreſly declares, That *GOD brought in the flood upon the world*, 2 Pet. 2. 5. Where (upon view of the Context) it will appear, that the Apoſtle makes the bringing in of the Flood, to be as much GOD's Work, as ever it was to caſt the ſinning Angels down to Hell, to ſave *Noah*; to burn *Sodom*; or to deliver *Lot*: all which were undeniably immediate and miraculous Acts of his. And truly that the Windows of Heaven ſhould be opened; and all the Fountains of the Great Deep broke up: that they ſhould be opened and broke up on the ſame day: that they ſhould be ſo opened and broke up, as to yield ſuch a quantity of Water at that time, as they never did before, and never did ſince, and never ſhall do again: what could this be but a ſpecial and wonderful Work of GOD?

I might farther obſerve the like miraculous workings of the DEITY, in *ſhutting up* thoſe Windows of Heaven again; and in *ſtopping* the aforeſaid Fountains of the Deep; and in *drying up* the Waters of the Deluge ſo faſt, *&c.* but I wave that (as I have done other things) to avoid prolixity.

Now when the Flood in all the periods of it, was thus disposed and govern'd by an Omnipotent and *Miraculous* hand; that the same hand should at once defend and direct the Ark; and so guard and steer it, as to keep it from Ship-wrack: is not at all to be wondred at. We may rather wonder, and wonder very much, if any should think otherwise.

To which add, That a miraculous protection and care of the Ark, would have been altogether as necessary, according to the *Theory*, or the *Old Hypothesis*. For, according to the *Theory*, the Ark must have sunk as low as the falling Earth; and then have been thrown up higher than the highest Mountains; and have been toss'd with such terrible and hideous jactations, as that the worst which are suffered on the roughest Seas, would scarce be shadows to them. So that unless a miraculous Providence superintended it, how could it be safe? And therefore indeed the *Theory* represents it, with its Guardian Angels about it, in the extremity of the Flood. And then according to the *Old way*, the tops of the Mountains must have been above Water, all the time that the Deluge was waxing. And so without such a Providence again, the Ark would have been as much imperill'd by those Mountains (if not more) as if they had been drown'd no deeper than *we* suppose them. Yea, in that very juncture when the Flood (according to the common account) was at its highest; the Ark *struck* upon the Mountains of *Ararat*, and was *stranded* there. And to save it in such circumstances, a most miraculous Providence was necessary indeed. But then the same may as lawfully be challenged *by*, and ought as readily to be allowed *to*, *our* Hypothesis likewise.

11. Which

11. Which grant; and then if in this memorable Flood, any difficulties be ſtarted, that Men are puzzled to make out: any *Phænomena's* ariſe, that are too big to proceed from *Nature* alone; and too intricate to be underſtood by *Reaſon*: lo, here's a *general* Anſwer to them, if not ſolution of them. The Flood was a *Miracle* in good meaſure. Or had ſo much miracle running through it, and interwoven with it; that all paſſages in it, are not to be accounted for by *Reaſon* and *Philoſophy*. And truly where *Nature* was over-ruled by *Providence*; it is but fit that *Philoſophy* ſhould give place to *Omnipotence*: and *Faith* ſway our Minds to aſſent to thoſe things, which *Reaſon* is unable to apprehend and explicate.

CHAP. XVII.

1. *The* Positiveness *of the* Theory. 2. *Noted in the* English Edition *of it*. 3. *Its Authors* Intentions *laudable*. 4. *The* Conclusion.

1. HAving gone over the several *Vital* or *Primary Assertions* of the *Theory*; I shall now only desire leave, briefly to note the *Positiveness* of it. It being indeed of an unusual Strain, and such as is seldom found in a new *Hypothesis*; especially at its first setting up, and sallying out into the World.

2. This *Positiveness* is very apparent, both in the *Latin*, and *English* Editions of the *Theory*. But I shall observe it only in the *latter*; that coming out *after* the other, and so with more deliberation and mature thoughts of things. It there discovers it self in such Passages as these:

I am willing to add here a Chapter or two, to shew that what we have delivered is *more than an Idea*, and that it was *in this very way* that *Noah*'s Deluge came to pass, *pag*. 79.

As we do not think it an unhappy discovery to have found out (*with a moral certainty*) the seat of the *Mosaical* Abyss,———so this gives us *a great assurance*, that the Theory we have given of a general Deluge, *is not a meer Idea*, but *is to be appropriated to the Deluge of* Noah, as *a true explication of it*, pag. 84.

That

That our Deſcription *is a reality*, both as to the Antediluvian Earth, and as to the Deluge, we may farther be *convinc'd* from St. *Peter's* Diſcourſe, *pag*. 85.

We may *ſafely conclude* that this *is no imaginary Idea*, but *a true account* of that ancient Flood whereof *Moſes* hath left us the Hiſtory, *ibid*.

If they (the ancient Earth and Abyſs) were in no *other form*, nor *other ſtate*, than what they are under now, the expreſſions of the ſacred Writers concerning them are *very ſtrange and inaccountable*; *without any ſufficient ground*, *or any juſt occaſion for ſuch uncouth repreſentations*. I fear there is ſomething more than Poſitiveneſs in this clauſe; which occurs, *pag*. 93.

We have *proved* our Explication of the Deluge to be *more than an Idea*, or to be *a true piece of Natural Hiſtory*; *and it may be the greateſt and moſt remarkable* that hath been ſince the beginning of the World. We have ſhown it to be *the real account* of *Noah's* Flood, *pag*. 96.

I confeſs, for my own part, when I obſerve how eaſily and naturally this *Hypotheſis* doth apply it ſelf to all the particularities of this Earth, hits and falls in ſo luckily and ſurprizingly with all the odd poſtures of its parts, *I cannot*, *without violence*, *bear off my mind from fully aſſenting to it*, pag. 113.

To ſpeak the truth, this Theory is *ſomething more than a bare* Hypotheſis, *pag*. 149.

It will never be beaten out of my head, but that St. *Peter* hath made the *ſame diſtinction* (we make of the Antediluvian Earth and Heavens from the Poſtdiluvian) ſixteen hundred years ſince, and *to the*

the very same purpose; so that *we have sure footing here again, and the* Theory *riseth above the Character of a bare* Hypothesis.——*We must in equity give more than a moral certitude to this* Theory, *pag.* 150.

I think there is nothing but the uncouthness of the thing to some Mens understandings, the custom of thinking otherwise, and the uneasiness of entring into a new sett of thoughts, that *can be a bar or hindrance* to its reception, *pag.* 170.

The *Theory* carries its *own light* and *proof* with it, *pag.* 274.

These are the Vitals of the *Theory*, and the Primary Assertions whereof I do *freely profess my full belief*, pag. 288.

Now I confess, I should have been much at a loss, whither to impute such extraordinary positive confidence, as shows it self (by these excerptions) in a Man so ingenious, touching things so precarious; had he not told me in this *Maxim* of his own: *A strong inclination, with a little evidence, is equivalent to a strong evidence*, pag. 297. Which considered; we need not wonder that strong Perswasions should sometimes be built upon weak grounds. Or to speak it in the *Theorist*s next words; *we are not to be surprised, if we find Men confident in their Opinions many times far beyond the degree of their evidence.*

3. Yet that his Intentions, in conposing and publishing his Book, were good and laudable; we have no reason to doubt. His own Declaration speaks them so. *I have no other design than to contribute my endeavours to find out the truth in a subject of so great*

great importance, and wherein the World hath hitherto had ſo little Satisfaction, pag. 97. A noble aim; but he that would cleverly hit the mark, muſt beware of ſhooting through Scripture, and wounding it at the rate the *Theoriſt* has done.

4. To Conclude. If ſo be, ſincere and upright Intentions will juſtifie the failures of a Pen, and in any meaſure ſerve to extenuate or excuſe them; I can take up that Plea in behalf of mine. And whereas in the new Explication of the Deluge, I may ſeem to have run out into a kind of *lax* interpretation of one or two Texts of Holy Scripture; I have ſufficiently apologiz'd for that excurſion already, by owning that (beſides it is neceſſary to expound thoſe Scriptures a new way, upon the account of the old *Hypotheſis* of the Flood) it was made but to vie with the *Theory*; and to try if we could hit upon another way of explaining the Deluge that might paſs for *rational* and intelligible. And therefore I only add this, which I do moſt heartily, I had rather, much rather my Papers ſhould be burnt to Aſhes, and my ſelf with them; than that I ſhould knowingly and wilfully write any thing, in way of oppoſition *to*, depravation *of*, or derogation *from*, any Divine Truth.

Οὐ δυνάμεθά τι κατὰ τῆς ἀληθείας, ἀλλ' ὑπὲρ τῆς ἀληθείας.

F I N I S.

History of Geology

An Arno Press Collection

Association of American Geologists and Naturalists. **Reports of the First, Second, and Third Meetings of the Association of American Geologists and Naturalists, at Philadelphia, in 1840 and 1841, and at Boston in 1842.** 1843

Bakewell, Robert. **An Introduction to Geology.** 1833

Buckland, William. **Reliquiae Diluvianae:** Or, Observations on the Organic Remains Contained in Caves, Fissures, and Diluvial Gravel. 1823

Clarke, John M[ason]. **James Hall of Albany:** Geologist and Palaeontologist, 1811-1898. 1923

Cleaveland, Parker. **An Elementary Treatise on Mineralogy and Geology.** 1816

Clinton, DeWitt. **An Introductory Discourse:** Delivered Before the Literary and Philosophical Society of New-York on the Fourth of May, 1814. 1815

Conybeare, W. D. and William Phillips. **Outlines of the Geology of England and Wales.** 1822

Cuvier, [Georges]. **Essay on the Theory of the Earth.** Translated by Robert Kerr. 1817

Davison, Charles. **The Founders of Seismology.** 1927

Gilbert, G[rove] K[arl]. **Report on the Geology of the Henry Mountains.** 1877

Greenough, G[eorge] B[ellas]. **A Critical Examination of the First Principles of Geology.** 1819

Hooke, Robert. **Lectures and Discourses of Earthquakes and Subterraneous Eruptions.** 1705

Kirwan, Richard. **Geological Essays.** 1799

Lambrecht, K. and W. and A. Quenstedt. **Palaeontologi:** Catalogus Bio-Bibliographicus. 1938

Lyell, Charles. **Charles Lyell on North American Geology.** Edited by Hubert C. Skinner. 1977

Lyell, Charles. **Travels in North America in the Years 1841-2.** Two vols. in one. 1845

Marcou, Jules. **Jules Marcou on the Taconic System in North America.** Edited by Hubert C. Skinner. 1977

Mariotte, [Edmé]. **The Motion of Water and Other Fluids.** Translated by J. T. Desaguliers. 1718

Merrill, George P., editor. **Contributions to a History of American State Geological and Natural History Surveys.** 1920

Miller, Hugh. **The Old Red Sandstone.** 1857

Moore, N[athaniel] F. **Ancient Mineralogy.** 1834

[Murray, John]. **A Comparative View of the Huttonian and Neptunian Systems of Geology.** 1802

Parkinson, James. **Organic Remains of a Former World.** Three vols. 1833

Phillips, John. **Memoirs of William Smith, LL.D.** 1844

Phillips, William. **An Outline of Mineralogy and Geology.** 1816

Ray, John. **Three Physico-Theological Discourses.** 1713

Scrope, G[eorge] Poulett. **The Geology and Extinct Volcanos of Central France.** 1858

Sherley, Thomas. **A Philosophical Essay.** 1672

Thomassy, [Marie Joseph] R[aymond]. **Géologie pratique de la Louisiane.** 1860

Warren, Erasmus. **Geologia:** Or a Discourse Concerning the Earth Before the Deluge. 1690

Webster, John. **Metallographia:** Or, an History of Metals. 1671

Whiston, William. **A New Theory of the Earth.** 1696

White, George W. **Essays on History of Geology.** 1977

Whitehurst, John. **An Inquiry into the Original State and Formation of the Earth.** 1786

Woodward, Horace B. **History of Geology.** 1911

Woodward, Horace B. **The History of the Geological Society of London.** 1907

Woodward, John. **An Essay Toward a Natural History of the Earth.** 1695